Spills of Diluted Bitumen from Pipelines

A Comparative Study of Environmental Fate, Effects, and Response

Committee on the Effects of Diluted Bitumen on the Environment

Board on Chemical Sciences and Technology

Division on Earth and Life Studies

The National Academies of
SCIENCES · ENGINEERING · MEDICINE

THE NATIONAL ACADEMIES PRESS
Washington, DC
www.nap.edu

THE NATIONAL ACADEMIES PRESS   500 Fifth Street, NW   Washington, DC 20001

This activity was supported by Contract No. DTPH5614C00001 with the U.S. Department of Transportation. Any opinions, findings, conclusions, or recommendations expressed in this publication do not necessarily reflect the views of any organization or agency that provided support for the project.

International Standard Book Number-13:   978-0-309-38010-2
International Standard Book Number-10:   0-309-38010-3
Library of Congress Control Number: 2015959867

Suggested citation: National Academies of Sciences, Engineering, and Medicine. 2016. *Spills of Diluted Bitumen from Pipelines: A Comparative Study of Environmental Fate, Effects, and Response.* Washington, DC: The National Academies Press.

COVER (*FRONT*):

(*top*)
Credit: John W. Poole/NPR
An oil sheen appears along the shore of the Kalamazoo River in August 2012. In July 2010, more than 800,000 gallons of tar sands oil entered Talmadge Creek and flowed into the Kalamazoo River, a Lake Michigan tributary. Heavy rains caused the river to overtop existing dams and carried oil 30 miles downstream.

(*bottom left*)
Credit: Jacqueline Michel
Sorbents and booms deployed in Dawson Cove in response to a crude oil spill in Mayflower, Arkansas in April 2013. In March 2013, over 3,000 barrels of crude oil spilled from a rupture in the Pegasus pipeline spilling oil in a residential neighborhood and eventually into a heavily wooded cove.

(*bottom center*)
Credit: Douglas Friedman
Oil spill response workers shuttling oiled debris from the beach below. In May 2015, an estimated 100,000 gallons of heavy crude oil discharged from Plains All American pipeline 901 near Refugio State Beach in Santa Barbara County, California.

(*bottom right*)
Credit: NOAA
On November 26, 2004, the single-hulled tanker Athos I unknowingly struck a large anchor submerged in the Delaware River while preparing to dock at a refinery just outside Philadelphia, Pennsylvania. The impact punctured the tanker's hull, and it began leaking more than 263,000 gallons of heavy oil into the tidal waters of this busy East Coast shipping route.
A worker in protective gear power-washes the oily rocks while boom in background collects oil five months after the spill occurred.

COVER (*BACK*):

Credit: Jonathon Gruenke
Jeremy Blackford of Clean Harbors uses a suction hose to clean oil from atop the Kalamazoo River in a containment area in Augusta, a village in Kalamazoo County in Michigan.

The National Academies of
SCIENCES · ENGINEERING · MEDICINE

The **National Academy of Sciences** was established in 1863 by an Act of Congress, signed by President Lincoln, as a private, nongovernmental institution to advise the nation on issues related to science and technology. Members are elected by their peers for outstanding contributions to research. Dr. Ralph J. Cicerone is president.

The **National Academy of Engineering** was established in 1964 under the charter of the National Academy of Sciences to bring the practices of engineering to advising the nation. Members are elected by their peers for extraordinary contributions to engineering. Dr. C. D. Mote, Jr., is president.

The **National Academy of Medicine** (formerly the Institute of Medicine) was established in 1970 under the charter of the National Academy of Sciences to advise the nation on medical and health issues. Members are elected by their peers for distinguished contributions to medicine and health. Dr. Victor J. Dzau is president.

The three Academies work together as the **National Academies of Sciences, Engineering, and Medicine** to provide independent, objective analysis and advice to the nation and conduct other activities to solve complex problems and inform public policy decisions. The Academies also encourage education and research, recognize outstanding contributions to knowledge, and increase public understanding in matters of science, engineering, and medicine.

Learn more about the National Academies of Sciences, Engineering, and Medicine at **www.national-academies.org**.

Preface

"I tell this story to illustrate the truth of the statement I heard long ago in the Army: *Plans are worthless, but planning is everything*. There is a very great distinction because when you are planning for an emergency you must start with this one thing: *the very definition of 'emergency' is that it is unexpected, therefore it is not going to happen the way you are planning*."

President Dwight D. Eisenhower
November 14, 1957

The transport of crude oil through transmission pipelines in the U.S. has been essential to move crude oil from production fields to refineries for many decades, and has thus been an integral aspect of the U.S. energy infrastructure. Starting with the impact of the large crude oil spill off the coast of Santa Barbara, California, in 1969, the inherent environmental risks associated with the transport of crude oil became more widely recognized. This spill contributed to the passage of the National Environmental Policy Act of 1969 (NEPA), the creation of the President's Council on Environmental Quality, and the U.S. Environmental Protection Agency. Several changes in the governmental approach to environmental policy would follow, eventually leading to the Oil Pollution Act of 1990 (OPA 90) in the wake of the *Exxon Valdez* spill in 1989. Through the resulting legislation of 1990 and Executive Order 12777, as amended, the U.S. Department of Transportation (USDOT) assumed responsibility to oversee the safe transport of crude oil in transmission pipelines, including thorough reviews of response plans and other actions.

Now, 25 years after OPA 90 was passed, a shift in the distribution of the types of crude oil carried in transmission pipelines has occurred and is anticipated to continue. Dense and viscous bitumen extracted using new technology from sources primarily in northern Alberta, Canada, is being diluted with less viscous hydrocarbons and transported to refineries throughout North America via transmission pipelines. This shift, along with a major spill of diluted bitumen in Marshall, Michigan, in 2010 and other spills elsewhere, has prompted Congress and USDOT to ask the National Academies of Sciences, Engineering, and Medicine to consider the use of transmission pipelines to transport diluted bitumen. The Academies' first study, released in 2013, focused on whether diluted bitumen was more likely to cause pipeline spills when compared to commonly transported crude oils. That study found no evidence of any causes of pipeline failure that are unique to the transportation of diluted bitumen. In this follow-on study, our committee was charged with addressing the question of whether the transport of diluted bitumen in pipelines has potential environmental consequences that are sufficiently different from those of commonly transported crude oils to warrant changes in regulations governing spill response planning, preparedness, and cleanup.

The committee brought together diverse expertise on the chemistry and environmental impacts of crude oils and broad experience in spill response. Two members, including the study chair, have backgrounds in hydrology and environmental engineering. We had members with expertise in oil chemistry, geochemistry and biogeochemistry, and oil fate, behavior, and toxicity. Several of these scientists have been, and continue to be, actively involved in oil spill response activities. Beyond the scientific and engineering expertise, experts in pipeline operations and environmental regulations ensured that the committee considered the practical and policy aspects of our recommendations.

In May 2015, while this study was still in its information-gathering phase, a rupture in Plains All American Line 901 spilled over 100,000 gallons of a heavy crude oil in Santa Barbara County, California and impacted almost 100 miles of shoreline. In addition to the two members of our committee who participated directly in the spill response as experts, we were able to observe the highly organized incident command in action four days after the spill. We discussed active response strategies with the National Oceanic and Atmospheric Administration (NOAA) Scientific Support Coordinator, the Liaison from the California Office of Spill Prevention and Response (CalOSPR), and several key members of the response team. A focus of these discussions was on the practical uses of formal response plans and on the daily decision-making process. As stated in President Eisenhower's famous quote, it was clear that the preparation of response plans was an invaluable process that has improved the effectiveness of response.

The focus of this report and its recommendations is on the current concerns related to the transport of diluted bitumen in pipelines. We are confident that, by updating the planning process and taking greater advantage of available information about diluted bitumen when it is spilled, the effectiveness of spill response can be enhanced. However, given the nature of pipeline operations, response planning, and the oil industry, it is likely that our recommendations will be applicable to spill response, preparedness, and cleanup for many types of crude oil.

Diane McKnight, Chair
Douglas Friedman, Study Director

Acknowledgments

The completion of this study would have not been successful without the assistance of many individuals and organizations. The committee would especially like to thank the following individuals and organizations for their contributions during this study:

U.S. Department of Transportation, Pipeline and Hazardous Materials Safety Administration, which sponsored the study and provided valuable information on the agency's responsibilities and structure. The committee would especially like to thank the Associate Administrator, Jeffrey Wiese, as well as Eddie Murphy, David Lehman, and Robert Smith. Mr. Smith served as the agency's liaison to the committee and was effective in responding to the committee's requests for information and site visits.

U.S. Coast Guard for providing information on the agency's regulatory responsibilities and technical information on the topic area. The committee would particularly like to thank Captain Claudia Gelzer, Captain Joseph Loring, Lt. Brandon Aten, and Lt. Sara Thompson.

U.S. Environmental Protection Agency, which provided information regarding the agency's regulatory responsibilities and experiences involved with oil spill response. The committee would like to thank Ralph Dollhopf, who served as an informal liaison to the committee; as well as Mark Howard, Greg Powell, Chris Ruhl, and Brian Schlieger.

Speakers and invited participants at the committee's data-gathering meetings. These individuals are listed here: Andy Black, Association of Oil Pipe Lines; Anthony Swift, Natural Resources Defense Council (NRDC); Bruce Hollebone, Environment Canada; Chris Reddy, Woods Hole Oceanographic Institution; Dan Capone, Mannik & Smith Group; David Westerholm, National Oceanic and Atmospheric Administration (NOAA); Faith Fitzpatrick, U.S. Geological Survey (USGS); Gary Shigenaka, NOAA; Heather Dettman, Natural Resources Canada; John Zhou, Alberta Innovates; Ken Lee, Centre for Offshore Oil, Gas and Energy Research (COOGER); Liam Stone, Government of Canada; Lyman Young; Paul Connors, Government of Canada; Peter Hodson, Queens University; Peter Lidiak, American Petroleum Institute (API); Robin Rorick, API; Steve Larter, University of Calgary; Steve Lehmann, NOAA; Thomas King, COOGER; Tim Nedwed, ExxonMobil; and Tom Miesner, Pipeline Knowledge and Development.

Jordan Stout, NOAA Scientific Support Coordinator, and **Joy Lavin-Jones**, Liaison Officer for California Office of Spill Prevention and Response (CalOSPR), for hosting a subgroup of the committee to observe the spill response operations for the Santa Barbara, California spill on May 19, 2015.

And last, but certainly not least, the **Academies staff** for organizing and facilitating this study. Study Director Douglas Friedman and Associate Program Officer Camly Tran organized the committee meetings and assisted the committee with research, report writing, and review. Senior Program Assistants Nawina Matshona and Cotilya Brown managed logistics of the meetings and publication. Senior Program Assistant Claire Ballweg and Communications Associate Sharon Martin contributed to the design of the figures and tables.

Acknowledgment of Reviewers

This report has been reviewed in draft form by individuals chosen for their diverse perspectives and technical expertise. The purpose of this independent review is to provide candid and critical comments that will assist the institution in making its published report as sound as possible and to ensure that the report meets institutional standards for objectivity, evidence, and responsiveness to the study charge. The review comments and draft manuscript remain confidential to protect the integrity of the deliberative process. We wish to thank the following individuals for their review of this report:

Mark Barteau, University of Michigan
Jim Elliott, T&T Marine Salvage, Inc.
Abbas Firoozabadi, Reservoir Engineering Research Institute
Katherine H. Freeman, The Pennsylvania State University
Elliott P. Laws, Crowell & Moring
Ken Lee, Commonwealth Scientific and Industrial Research
    Organization
Patricia Maurice, University of Notre Dame
Stephen A. Owens, Squire Patton Boggs
Chris Reddy, Woods Hole Oceanographic Institution
Calvin H. Ward, Rice University

Although the reviewers listed above have provided many constructive comments and suggestions, they were not asked to endorse the conclusions or recommendations nor did they see the final draft of the report before its release. The review of this report was overseen by **Thomas Leschine** of the University of Washington and **Michael Ladisch** of Purdue University, who were responsible for making certain that an independent examination of this report was carried out in accordance with institutional procedures and that all review comments were carefully considered. Responsibility for the final content of this report rests entirely with the authoring committee and the institution.

Acronyms

ACP	Area Contingency Plan
ADIOS	Automated Data Inquiry for Oil Spills
AOPL	Association of Oil Pipe Lines
API	American Petroleum Institute
AWB	Access Western Blend
BSEE	Bureau of Safety and Environmental Enforcement
BTEX	benzene, toluene, ethylbenzene, xylenes
CalOSPR	California Office of Spill Prevention and Response
CERCLA	Comprehensive Environmental Response, Compensation, and Liability Act
CLWB	Cold Lake Winter Blend
COOGER	Centre for Offshore Oil, Gas and Energy Research
Da	Dalton
DSD	droplet size distribution
FRP	Facility Response Plan
GNOME	General NOAA Operational Modeling Environment
HCA	high-consequence area

ICCOPR	Interagency Coordinating Committee on Oil Pollution Research
MCL	maximum contaminant level
NAPL	non-aqueous-phase liquid oil
NAS	National Academy of Sciences
NCP	National Contingency Plan
NEB	National Energy Board
NEPA	National Environmental Policy Act
NGO	nongovernmental organization
NOAA	National Oceanic and Atmospheric Administration
NOSAMS	National Ocean Sciences Accelerator Mass Spectrometry
NRC	National Research Council
NRDA	Natural Resource Damage Assessment
NRDC	Natural Resources Defense Council
NRT	National Response Team
OPA 90	Oil Pollution Act of 1990
OPA	oil-particle aggregate
OSC	On-Scene Coordinator
OSHA	Occupational Safety and Health Administration
OSPR	Office of Spill Prevention and Response
OSRO	Oil Spill Removal Organization
PAH	polycyclic aromatic hydrocarbon
PHMSA	Pipeline and Hazardous Materials Safety Administration
RRT	Regional Response Team
SDWA	Safe Drinking Water Act
SDS	Safety Data Sheet
TM	Trans Mountain
TPH	Total Petroleum Hydrocarbons
USCG	U.S. Coast Guard
USDOT	U.S. Department of Transportation
USEPA	U.S. Environmental Protection Agency
USGS	U.S. Geological Survey
VOC	volatile organic compound

Contents

Summary

In January 2012, Congress tasked the Secretary of Transportation to "determine whether any increase in the risk of release exists for pipelines transporting diluted bitumen."[1] In response to the congressional request, the U.S. Department of Transportation (USDOT) asked the National Academies of Sciences, Engineering, and Medicine (the Academies) to study the likelihood of release of diluted bitumen from crude oil transmission pipelines. The Academies released a report in 2013 concluding that "[t]he committee does not find any causes of pipeline failure unique to the transportation of diluted bitumen."[2] Following the 2013 release of *Effects of Diluted Bitumen on Crude Oil Transmission Pipelines*, Congress subsequently charged USDOT to "investigate whether the spill properties [of diluted bitumen] differ sufficiently from other liquid petroleum products to warrant modifications to the spill response plans, spill preparedness, or cleanup regulations and report on those findings to the House and Senate Committees on Appropriations within 180 days of enactment."[3]

USDOT returned to the Academies in 2014 with a request to form an ad hoc committee to help address this concern. Specifically, this committee was tasked[i] to review the available literature and data to examine the current state of knowledge, and to identify the relevant properties and characteristics of the transport, fate, and effects of diluted bitumen and commonly transported crude oils when spilled in the environment from U.S. transmission pipelines. Based on a comparison of the relevant

---

[i] The committee's full statement of task can be found in Box 1-1.

properties of diluted bitumen and of a representative set of crude oils that are commonly transported via pipeline, the committee was asked to determine whether the differences between properties of diluted bitumen and those of other commonly transported crude oils warrant modifications to the regulations governing spill response plans, preparedness, and cleanup.

STUDY APPROACH

In order to answer the questions outlined in the statement of task, the committee analyzed information in a variety of forms. Part of the committee's data gathering included hearing presentations, meeting with stakeholders, and reviewing the literature. A detailed list of the individuals the committee met with can be found in the Acknowledgments section of this report. In the early phases of the study, an opportunity for public comment was provided. After considering all of the available data and information, the history of the study, and the sponsor's request, the committee focused on environments that would most likely be affected by an oil spill from a transmission pipeline–that is, the contiguous U.S., including the near-shore coastline with far offshore not being considered. The report also focuses on spills from transmission pipelines and does not explicitly address other modes of transportation (e.g., rail, barge, truck, and tanker). It is likely that many of the topics covered in this report, and many of the conclusions and recommendations, will be applicable to these other transportation modes because many aspects of environmental impact are independent of mode of transportation.

The committee's task requires a comparison between diluted bitumen and "crude oils commonly transported in U.S. transmission pipelines." After an analysis of the total volumes of crude oil transported by U.S. pipelines (see Chapter 1), a set of light and medium crudes was chosen as representative of those "commonly transported" and likely to be encountered in a response scenario. The committee's approach is described in greater detail in Chapter 1.

KEY FINDINGS AND CONCLUSIONS

The starting point for assessing the **Chemical and Physical Properties of Crude Oils (Chapter 2)** was the intrinsic complexity of crude oils as mixtures of hydrocarbons with diverse structures and widely varying molecular weights. Mixtures of these compounds combine to make up the bulk properties of any particular crude oil. The bitumen fraction, in particular, is associated with reservoirs of recalcitrant and immobile crude oils. Unconventional extraction methods are required to access bitumen reservoirs and addition of a diluent is needed to transport the bitumen product through unheated transmission oil pipelines. In comparison to other commonly transported crude oils, many of the chemical and physical properties of diluted bitumen, especially those relevant to environmental impacts, are found to differ substantially from those of the other crude oils. *The key differences are in the exceptionally high density, viscosity, and adhesion properties of the bitumen component of the diluted bitumen that dictate environmental behavior as the crude oil is subjected to weathering (a term that refers to physical and chemical changes of spilled oil).*

Immediately following a spill, the **Environmental Processes, Behavior, and Toxicity of Diluted Bitumen (Chapter 3)** are similar to those of other commonly transported crudes. Beginning immediately after a spill, however, exposure to the environment begins to change spilled diluted bitumen through various weathering processes. The net effect is a reversion toward properties of the initial bitumen. An important factor is the amount of time necessary for the oil to weather into an adhesive, dense, viscous material. For any crude oil spill, lighter, volatile compounds begin to evaporate promptly; in the case of diluted bitumen, a dense, viscous material with a strong tendency to adhere to surfaces begins to form as a residue. *For this reason, spills of diluted bitumen pose particular challenges when they reach water bodies. In some cases, the residues can submerge or sink to the bottom of the water body.* Importantly, the density of the residual oil does not necessarily need to reach or exceed the density of the surrounding water for this to occur. The crude oil may combine with particles present in the water column to submerge, and then remain in suspension or sink.

These factors are important to consider for **Spill Response Planning and Implementation (Chapter 4)**. Spills of diluted bitumen into a body of water initially float and spread while evaporation of volatile compounds may present health and explosion hazards, as occurs with nearly all crude oils. It is the subsequent weathering effects, unique to diluted bitumen, that merit special response strategies and tactics. For example, the time windows during which dispersants and in situ burning can be used effectively are significantly shorter for diluted bitumen than for other commonly transported crudes. In cases where traditional removal or containment techniques are not immediately successful, the possibility of submerged and sunken oil increases. *This situation is highly problematic for spill response because (1) there are few effective techniques for detection, containment, and recovery of oil that is submerged in the water column, and (2) available techniques for responding to oil that has sunk to the bottom have variable effectiveness depending on the spill conditions.*

When **Comparing Properties Affecting Transport, Fate, Effects, and Response (Chapter 5)**, several key properties emerge. Figure S-1 illustrates the properties relevant to transport, fate, and effects and the potential environmental outcomes following a crude oil spill. Based on the similarities and differences between diluted bitumen (in pipeline and weathered forms) and other commonly transported crudes, the comparative levels of concern associated with these properties are highlighted. *The majority of the properties and outcomes that differ from commonly transported crudes are associated not with freshly spilled diluted bitumen, but with the weathering products that form within days after a spill.* Given these greater levels of concern for weathered diluted bitumen, spills of diluted bitumen should elicit unique, immediate actions in response.

Based on the differences identified previously, a review of the **Regulations Governing Spill Response Planning (Chapter 6)** was conducted. Of particular focus was Part 194 of the Pipeline and Hazardous Materials Safety Administration (PHMSA) regulations, which governs the planning of responses to spills from transmission pipelines. In addition, because the scope of the task was broadly defined to address "regulations governing spill response plans, spill preparedness, or cleanup," relevant U.S. Environmental Protection Agency (USEPA) and U.S. Coast Guard (USCG) regulations were reviewed, primarily for comparison to PHMSA regulations. It is clear that PHMSA takes a substantially different approach from USEPA and USCG when setting expectations for and reviewing spill response plans. Notably, PHMSA reviews plans for completeness in terms of the regulatory requirements only, while USEPA and USCG review plans for both completeness and adequacy for response. *Broadly, regulations and agency practices do not take the unique properties of diluted bitumen into account, nor do they encourage effective planning for spills of diluted bitumen.*

In light of the aforementioned analysis, comparisons, and review of the regulations, it is clear that **the differences in the chemical and physical properties relevant to environmental impact warrant modifications to the regulations governing diluted bitumen spill response plans, preparedness, and cleanup.** The concern associated with these differences is summarized in Figure S-1 for both diluted bitumen and weathered diluted bitumen. Each property that is relevant to environmental transport, fate, and effects is identified with the potential outcomes and a qualitative level of concern compared to other commonly transported crudes. The most notable changes observed are in the comparison between diluted bitumen and.weathered diluted bitumen. For example, the level of concern goes from the same to more (or less) concern between the weathered and non-weathered material for ten of the properties in Figure S-1 and all techniques identified in Figure S-2.

	Property	Potential Outcomes	Level of Concern Relative to Commonly Transported Crude Oils	
			Diluted Bitumen	Weathered Diluted Bitumen
Transport	Density	• Movement in suspension or as bedload	SAME	MORE
	Adhesion	• Movement in suspension or as bedload (oil particle aggregates)	MORE	MORE
	Viscosity	• Movement as droplets • Spreading on land • Groundwater contamination	SAME	LESS
	Solubility	• Mobility and toxicity in water	SAME	LESS
	BTEX	• Toxicity (water and air emissions)	LESS	LESS
Fate	Density	• Sinking • Burial	SAME	MORE
	Adhesion	• Sinking after sediment interaction • Surface coating	SAME	MORE
	Viscosity	• Penetration	LESS	LESS
	Percentage of light fraction	• Air emissions	SAME	LESS
	Flammability	• Fire or explosion risk	SAME	LESS
	Biodegradability	• Persistence	MORE	MORE
	Burn residue	• Quantity of residue • Residue sinking	MORE	MORE
Effects	Density	• Impaired water quality from oil in the water column and sheening	SAME	MORE
	Adhesion	• Fouling and coating	MORE	MORE
	BTEX components	• Contaminated drinking water • Respiratory problems/disease	SAME	LESS
	HMW components	• Trophic transfer/food web • Aquatic toxicity	UNKNOWN	
	LMW components	• Aquatic toxicity	UNKNOWN	
		• Taste/odor concerns in drinking water	SAME	LESS

The relative level of concern for diluted bitumen is

Less   Same   More

when compared to commonly transported crudes.

FIGURE S-1 Spill hazards: diluted bitumen relative to commonly transported crude oils. Acronyms: BTEX: benzene, toluene, ethylbenzene, xylenes; HMW: high molecular weight; LMW: low molecular weight.

	Technique	Potential Outcomes	Level of Concern Relative to Commonly Transported Crude Oils	
			Diluted Bitumen	Weathered Diluted Bitumen
Response Operations	Worker/public safety from explosion risk/ VOCs	• Public evacuation • Worker respiratory protection/personal safety	SAME	LESS
	Booming/ skimming	• More difficult due to changes in viscosity/density	SAME	MORE
	In situ burning	• Narrow window of opportunity/residue sinking	MORE	MORE
	Dispersants	• Narrow window of opportunity	MORE	MORE
	Surface cleaning agents	• More aggressive removal to meet cleanup endpoints	MORE	MORE
	Submerged/ sunken oil detection/ recovery	• More complex response • Less effective recovery for submerged/sunken oil	SAME	MORE
	Waste generation	• Higher removal volumes from residue persistence	MORE	MORE
		• Sunken oil recovery	SAME	

The relative level of concern for diluted bitumen is

    Less   Same   More

when compared to commonly transported crudes.

FIGURE S-2 Response operations: diluted bitumen relative to commonly transported crude oils. Acronym: VOCs: volatile organic compounds.

RECOMMENDATIONS

Diluted bitumen has unique properties, differing from those of commonly transported crude oils, which affect the behavior of diluted bitumen in the environment following a spill. This behavior differs from that of the light and medium crudes typically considered when planning responses to spills. Of greatest significance are the physical and chemical changes that diluted bitumen undergoes during weathering. A more comprehensive and focused approach to diluted bitumen across the oil industry and the relevant federal agencies is necessary to improve preparedness for spills of diluted bitumen and to spur more effective cleanup and mitigation measures when these spills occur. The recommendations presented here are designed to achieve this goal.

Oil Spill Response Planning

Recommendation 1: To strengthen the preparedness for pipeline releases of oil from pipelines, the Part 194 regulations implemented by PHMSA should be modified so that spill response plans are effective in anticipating and ensuring an adequate response to spills of diluted bitumen. These modifications should

a. Require the plan to identify all of the transported crude oils using industry-standard names, such as Cold Lake Blend, and to include safety data sheets for each of the named crude oils. Both the plan and the associated safety data sheets should include spill-relevant properties and considerations;

b. Require that plans adequately describe the areas most sensitive to the effects of a diluted bitumen spill, including the water bodies potentially at risk;

c. Require that plans describe in sufficient detail response activities and resources to mitigate the impacts of spills of diluted bitumen, including capabilities for detection, containment, and recovery of submerged and sunken oil;

d. Require that PHMSA consult with USEPA and/or USCG to obtain their input on whether response plans are adequate for spills of diluted bitumen;

e. Require that PHMSA conduct reviews of both the completeness and the adequacy of spill response plans for pipelines carrying diluted bitumen;

f. Require operators to provide to PHMSA, and to make publicly available on their websites, annual reports that indicate the volumes of diluted bitumen, light, medium, heavy, and any other crude oils carried by individual pipelines and the pipeline sections transporting them; and

g. Require that plans specify procedures by which the pipeline operator will (i) identify the source and industry-standard name of any spilled diluted bitumen to a designated Federal On-Scene Coordinator, or equivalent state official, within 6 hours after detection of a spill and (ii) if requested, provide a 1-L sample drawn from the batch of oil spilled within 24 hours of the spill, together with specific compositional information on the diluent.

Oil Spill Response

Recommendation 2: USEPA, USCG, and the oil and pipeline industry should support the development of effective techniques for detection, containment, and recovery of submerged and sunken oils in aquatic environments.

Recommendation 3: USEPA, USCG, and state and local governments should adopt the use of industry-standard names for crude oils, including diluted bitumen, in their oversight of oil spill response planning.

USCG Classification System

Recommendation 4: USCG should revise its oil-grouping classifications to more accurately reflect the properties of diluted bitumen and to recognize it as a potentially nonfloating oil after evaporation of the diluent. PHMSA and USEPA should incorporate these revisions into their planning and regulations.

Advanced Predictive Modeling

Recommendation 5: NOAA should lead an effort to acquire all data that are relevant to advanced predictive modeling for spills of diluted bitumen being transported by pipeline.

Improved Coordination

Recommendation 6: USEPA, USCG, PHMSA, and state and local governments should increase coordination and share lessons learned to improve the area contingency planning process and to strengthen preparedness for spills of diluted bitumen. These agencies should jointly conduct announced and unannounced exercises for spills of diluted bitumen.

Improved Understanding of Adhesion

Recommendation 7: USEPA should develop a standard for quantifying and reporting adhesion because it is a key property of fresh and weathered diluted bitumen. The procedure should be compatible with the quantity of the custodial sample collected by pipeline operators.

RESEARCH PRIORITIES

Although many differences between diluted bitumen and commonly transported crudes are well established, there remain areas of uncertainty that hamper effective spill response planning and response to spills. These uncertainties span a range of issues, including diluted bitumen's behavior in the environment under different conditions, its detection when submerged or sunken, and the best response strategies for mitigating the impacts of submerged and sunken oil. These research priorities, discussed in Chapter 7, apply broadly to the research community.

Major topics for future research include

- *Transport and fate in the environment,*
- *Ecological and human health risks of weathered diluted bitumen,*
- *Detection and quantification of submerged and sunken oil,*
- *Techniques to intercept and recover submerged oil on the move, and*
- *Alternatives to dredging to recover sunken oil.*

1

Introduction

BACKGROUND

Over the past decade, production of "unconventional" oil in North America has surged as technological improvements and cost reductions have made these crudes competitive in the North American market. Unconventional oil in North America derives from two sources. In the U.S., hydraulic fracturing technologies have been widely applied to extract oil from shale formations or other typically inaccessible, low-permeability rocks. In Canada, petroleum products have been extracted from "oil sands" or "tar sands." Together, these streams have increased North American production of crude oils by 46% since 2008.[4]

The oil sands yield bitumen, a highly viscous form of petroleum that is produced by surface mining or by in situ recovery. Surface mining is preferred for deposits within 75 m of the surface.[5] In situ recovery, in which steam is injected to mobilize bitumen underground, is used for deeper deposits. Production of bitumen is predicted to increase 2.5-fold from 2013 to 2030[5a] although future production trends may be influenced by declining crude prices in world markets.

After separation from the host rock, bitumen is modified for transport. Commonly, it is combined with lower-density hydrocarbon mixtures (condensates, synthetic crude, or a mixture of both) to obtain a product with an acceptable viscosity and density for transport to refineries via pipeline. This engineered fluid is referred to as diluted bitumen. Common names refer to subtypes (e.g., dilbit, synbit, railbit, and dilsynbit) but, for

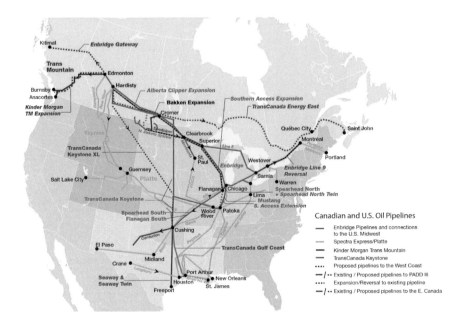

FIGURE 1-1 Existing and proposed Canadian and U.S. crude oil pipelines.
SOURCE: Canadian Association of Petroleum Producers[5a]

simplicity, the term diluted bitumen as used in this report encompasses all bitumen blends that have been mixed with lighter products.

Diluted bitumen has been transported by pipeline in the U.S. for more than 40 years, with the amount increasing recently as a result of improved extraction technologies and resulting increases in production and exportation of Canadian diluted bitumen. The increased importation of Canadian diluted bitumen to the United States has strained the existing pipeline capacity and contributed to the expansion of pipeline mileage over the past 5 years. Although rising North American crude production has resulted in greater transport of crude oil by rail or tanker, oil pipelines continue to deliver the vast majority of crude oil supplies to U.S. refineries. Most of the pipeline systems that are currently transporting diluted bitumen originate near extraction sites in Alberta, Canada (see Figure 1-1). To accommodate increased export volumes, additional pipelines are being proposed and developed. Proposals include (i) the Keystone XL.,[i] which would deliver diluted bitumen to Cushing, Oklahoma, in the U.S.; (ii)

---

[i] President Obama announced on November 6, 2015 the decision to deny a Presidential Permit for the construction of the Keystone XL Pipeline.

the Energy East, which would transport products to eastern Canada and its refineries; (iii) the Northern Gateway ("Enbridge Gateway" in Figure 1-1); and (iv) the Kinder Morgan Trans Mountain ("TM Expansion" in Figure 1-1). Both of the latter pipelines would transport products within Canada from Alberta to West Coast terminals.

In the event of a spill, impacts and cleanup procedures depend strongly on the environmental setting. Figure 1-1 indicates that the transmission pipelines transporting diluted bitumen are currently located onshore, which includes passage across terrestrial and freshwater environments, and near shore, which includes the marine waters near the coastline. Deepwater environments are not presently pertinent to pipeline transport of diluted bitumen and are not considered in this report.

EFFECTS OF DILUTED BITUMEN ON CRUDE
OIL TRANSMISSION PIPELINES

In January 2012, the Secretary of Transportation was tasked by Congress to "determine whether the regulations are sufficient to regulate pipeline facilities used for the transportation of diluted bitumen . . . and whether any increase in the risk of release exists for pipeline facilities transporting diluted bitumen."[2] The U.S. Department of Transportation's (USDOT) Pipeline and Hazardous Materials Safety Administration (PHMSA) contracted with the National Academies of Sciences, Engineering, and Medicine (the Academies) to assemble a committee of experts to analyze whether the likelihood of release was greater for the transportation of diluted bitumen compared to that for other commonly transported crudes via U.S. transmission pipelines.[2] An expert committee completed a comprehensive analysis and review of the available data on the chemical and physical properties of shipments of diluted bitumen and other crudes, examined pipeline incident statistics and investigations, and consulted experts in pipeline corrosion, cracking, and other causes of releases. The analysis covered many aspects of pipeline transportation including an explanation of the U.S. pipeline system; pipeline construction, maintenance, and alerts; incident data reported to PHMSA; and a discussion of bitumen production. Ultimately, after detailed analysis, the committee report, issued in 2013, "did not find any causes of pipeline failure unique to the transportation of diluted bitumen."[2] Environmental consequences of spills of diluted bitumen from pipelines were not within the scope of the *Effects of Diluted Bitumen on Crude Oil Transmission Pipelines* report. Following the release of the 2013 study, Congress tasked USDOT in 2014 to undertake a study to better understand the environmental impacts of spills of diluted bitumen from transmission pipelines and the adequacy of spill response planning.[3]

CHARGE TO THE COMMITTEE

Based on this direction from Congress, PHMSA returned to the Academies with a request to assemble an ad hoc committee to analyze whether the relevant properties of diluted bitumen differ sufficiently from those of other crude oils commonly transported in U.S. transmission pipelines to warrant modifications of the regulations governing spill response plans, spill preparedness, and/or cleanup. The committee's statement of task is provided in Box 1-1. This report focuses primarily on spills of crude oil from U.S. transmission pipelines. Over the course of producing this report, however, it became clear that the utilization of other modes of transportation for crude oil such as rail, truck, and tanker have increased and are worth consideration. While this report does not address any of the particular aspects of those transportation modes, many of the environmental effects of spilled oil are independent of the method of transporta-

---

BOX 1-1
Statement of Task

An ad hoc committee will analyze whether the properties of diluted bitumen differ sufficiently from those of other crude oils commonly transported in U.S. transmission pipelines to warrant modifications of the regulations governing spill response plans, spill preparedness, or cleanup.

The committee will
1. Review the available literature and data, including any available data from oil spill responses or cleanup, to determine the current state of knowledge of the transport, fate, and effects of diluted bitumen once spilled into the environment (onshore and offshore);
2. Identify the relevant properties and characteristics that influence the transport, fate, and effects of commonly transported crude oils, including diluted bitumen, in the environment;
3. Make a comparison of the relevant properties identified in item 2 between diluted bitumen and a representative set of crude oils that are commonly transported via pipeline; and
4. Based on the comparison in item 3, analyze and make a determination as to whether the differences between the environmental properties of diluted bitumen and those of other crude oils warrant modifications to the regulations governing spill response plans, spill preparedness, or cleanup.

If the committee finds that there is not sufficient information to make a comparison of the environmental properties between diluted bitumen and other crude oils, the committee may make recommendations as to the additional data that would be needed to make such a determination.

tion and therefore this report can provide useful insight into areas beyond pipeline transportation.

ADDRESSING THE STATEMENT OF TASK

To understand the potential consequences of spills of diluted bitumen, knowledge regarding its chemical properties and environmental behavior during and after a spill in various spill environments is required. To date, several reports have been published that examine the properties,[6] toxicity,[7] and composition of diluted bitumen products derived from the Canadian oil sands.[8] Other recent reports focus on the behavior and fate of spills in marine[5b,8-9] and freshwater environments.[5b,9a,10] Many of these reports were prepared as a result of the release of diluted bitumen into the Kalamazoo River by a break in the Enbridge 6B pipeline in Marshall, Michigan, on July 25, 2010. The total release was estimated to be 843,444 gallons, one of the largest freshwater oil spills in North American history, with cleanup costs exceeding $1.2 billion.[11] The Marshall release attracted attention because of the broad extent and consequences of the release and the unprecedented scale of impact.[12] The data and information gathered from experts and reports have been critical to addressing the statement of task and supporting the recommendations found herein. Nonetheless, the current knowledge base is limited, and a better understanding of the chemical constituents and behavior of diluted bitumen spills in diverse environmental settings would be helpful to inform response plans and actions.

Data Gathering

To make a comparative analysis of diluted bitumen and crudes commonly transported by U.S. transmission pipelines, the committee gathered information from a variety of experts and stakeholders from government, nongovernmental organizations (NGOs), industry, and academia. A list of those experts and stakeholders can be found in the Acknowledgments. Technical information on properties of crude oil and on the behavior, fate, and environmental impacts of spills of diluted bitumen was provided directly or presented during one of several data-gathering meetings. In addition, discussions with agencies, individuals, and groups concerned with development of plans for responses to oil spills were extremely valuable. A subgroup from the committee conducted a site visit to the incident command post for the Refugio spill in Santa Barbara, California, in May 2015 to observe a response in action and to hold discussions with participants. A questionnaire requesting data was submitted to the American Petroleum Institute (API) and to the Association of Oil Pipe Lines (AOPL) with a request for help obtaining data from the pipeline industry but drew no response.

Defining Commonly Transported Crude
Oil in the U.S. Pipeline System

In 2013, the United States produced 2,720 million barrels[13] of crude oil, an increase of about 35% since 2000.[14] The United States also imported a total of 2,820 million barrels[13] of crude oil in 2013 from all countries. The total volume of produced and imported crude oil to the United States for 2013, the most recent year for which complete data are available, was thus 5,540 million barrels. Of these, 3,190 million barrels[14] (58%) were delivered to refineries by pipeline. This amount includes both domestic crude (2,200 million barrels) and imported crude (992 million barrels).[14] A summary of products transported in the U.S. pipeline system is presented in Table 1-1. Taken in broad strokes, the majority of the crude oil transported in the U.S. pipeline system in 2013 was conventional light and medium crude (~71%). With the recently increased production of light crude oil in the U.S., it is expected that light and medium crudes will remain dominant in the U.S. pipeline system.

In the United States, production of heavy crude oil has been roughly constant even as production of light crude oil production has, in recent years, grown rapidly.[15] The principal domestic sources of heavy crude oil are in California, which produced 199 million barrels in 2013,[16] most of that being transported by pipeline within the state.[17] Because heated

TABLE 1-1 Types and Quantities of Crude Oil in the U.S. Pipeline System in 2013

Type of Crude Oil	Volume in 2013 (million barrels)	% by volume in pipeline system
Diluted Bitumen	250	8%
Undiluted Conventional Heavy[a]	199	6%
Diluted Conventional Heavy[b]	273	9%
Conventional Medium and Light	2,278[c]	71%
Synthetic Crude	190	6%
Total	3,190	

[a]Domestic production of conventional heavy crude oil, API < 27°.
[b]Imported conventional heavy crude oil, API < 25°.
[c]Conventional medium and light crude oil = Reported Total – (Diluted Bitumen + Diluted and Undiluted Conventional Heavy + Synthetic Crude).
SOURCES: National Energy Board and U.S. Energy Information Administration[15, 19]

pipelines are used, the heavy crude oils in California do not require blending with lighter products and are hence termed "undiluted conventional heavy" crude oil. Most of the other heavy crude oil that is transported by pipeline in the United States is from Canada. These heavy crude oils are diluted with lighter hydrocarbons and, hence, are referred to as "diluted conventional heavy" crude oil. Some imported heavy crude oil also comes from Mexico and Venezuela but those products arrive by tanker[18] and are not typically transported by U.S. transmission pipelines.

The remaining categories of crude oil transported via pipeline are diluted bitumen and synthetic crude. In 2013, the National Energy Board (NEB) of Canada reported an export of 250 million barrels of diluted bitumen to the United States.[20] The NEB has defined diluted bitumen as bitumen blended with light hydrocarbons and/or synthetic crude oil. Although there has been an increase in rail transportation of diluted bitumen, petroleum products from Canada, including diluted bitumen, are transported mainly by pipeline. By this analysis, diluted bitumen made up 8% of the crude oil carried in the U.S. pipeline system in 2013.[20] The volume of diluted bitumen imported from Canada increased by ~20% in 2014. The Canadian diluted bitumen transported in transmission pipelines to the U.S. typically contains 50-70% bitumen by volume with lighter hydrocarbons accounting for the remainder.[2,5b] The quantity of diluents added is typically the minimum needed to meet pipeline specifications. The most common specifications for pipeline inputs are a maximum density of 0.94 grams per cubic centimeter (g/cm^3) and a maximum viscosity of 350 centistokes (cSt).[2,6] Bitumen blended with synthetic crude usually has a mixture of about 50% bitumen and 50% synthetic crude,[5b] whereas bitumen blended with naphtha-based oils derived from conventional crudes or from condensates derived from natural gas typically contains a mixture of about 70% bitumen and 30% light oils.[2] Bitumen blends also vary seasonally in order to meet specifications for density and viscosity at the temperature of the pipeline.

Synthetic crude oil can be upgraded bitumen, upgraded heavy crude oil, or a mixture of those products, and makes up 6% of the total oil being transported by pipeline. "Upgrading" refers to inefficient, but cost-effective, refining procedures implemented at or near the site of production rather than after transport to a refinery. The total volumes of Canadian crudes imported to the U.S. by pipeline are presented in Table 1-2.

The statement of task seeks a comparison between diluted bitumen and "crude oils commonly transported in U.S. transmission pipelines." A definition of commonly transported crudes is thus required. Figure 1-2 graphically depicts the volume percentages of types of crude oil transported by the U.S. pipeline system in 2013 and includes both imported and domestic oil. It shows that light and medium crude oils are the

TABLE 1-2 Volumes of Canadian Crudes Imported to the U.S. by Pipeline in 2013

Type of Crude Oil	Volume (million barrels)
Diluted Bitumen[a]	250
Conventional Diluted Heavy[b]	273
Conventional Medium and Light	206
Synthetic Crude[c]	190

[a]Bitumen blended with light hydrocarbons and/or synthetic crude oil.
[b]API gravity < 25°.
[c]Upgraded bitumen or upgraded heavy crude oil of any API gravity.
SOURCE: National Energy Board[19]

predominant crude oil products being transported in the U.S. transmission pipeline system and account for nearly three-quarters of the crude oil transported. Further, a significant fraction of the transport of heavy crude oils occurs in a single state (California), whereas other crude oils are transported throughout the contiguous United States. This is the basis for identifying light and medium crudes as commonly transported and indicating that a comparison between diluted bitumen and these crude classes provides a meaningful basis for addressing the statement of task. Accordingly, for the purposes of this report, commonly transported crudes are defined as conventional light and medium crude oils (Figure 1-2).

Key terms used to describe the types of crude oils used throughout this report are highlighted in Figure 1-3.

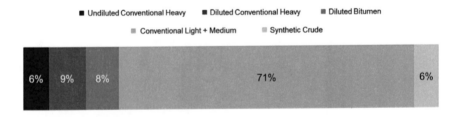

FIGURE 1-2 Percentages of crude oil types by volume in the U.S. pipeline system in 2013. See Table 1-1 for details.

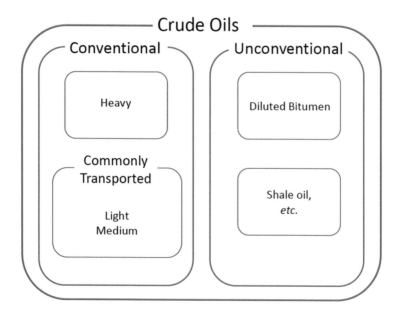

FIGURE 1-3 Key terms used in this report.

ORGANIZATION OF THE REPORT

Chapter 2 discusses the chemical and physical properties of crude oils that are relevant to environmental impact. Properties discussed in detail include density, viscosity, flash point, and adhesion. The effects of weathering on these properties are also highlighted and presented in a series of tables which are organized to compare light, medium, and heavy crude oils with diluted bitumen. The chapter concludes by identifying key differences between the properties of those products.

Chapter 3 examines the environmental transport, fate, and effects of spills of crude oils with a focus on properties unique to diluted bitumen before and after the diluent has been lost to volatilization. It also reviews relevant crude oil spills and considers potential spills in a variety of environmental settings including land, groundwater, inland waters, and coastal zones.

Chapter 4 describes the current planning and implementation of response to spills of crude oil. Each spill is unique and its characteristics depend on the chemical and physical properties of the oil and on the environment in which the spill has occurred. Predictions of the behavior of spilled oil and of its effects on health and safety are described in this chapter and are pertinent to how a spill response will be implemented and what type of tools and equipment will be employed. The chapter reviews general response tactics and techniques for floating oils as well as tactics for detection, containment, and recovery of spills that have a higher tendency to submerge. Considering the distinctions between light and medium crudes compared to that of diluted bitumen described in Chapters 2 and 3, a descriptive table about recovery techniques for diluted bitumen spills concludes the chapter.

Chapter 5 synthesizes the information presented in the previous chapters. The differences between commonly transported crude oils and diluted bitumen are presented in three separate tables organized in terms of environmental transport, fate, effects, and spill response describing the relevant properties, potential outcomes, and levels of concern.

Chapter 6 focuses on the adequacy of current regulations governing spill response plans, preparedness, and cleanup. The chapter provides an overview of the federal spill planning and response framework for crude oil spills. Weaknesses of the current pipeline spill response planning and response framework for addressing spills of diluted bitumen are discussed.

Chapter 7 presents specific recommendations to stakeholders involved in spill response based on the committee's analysis and assessment of the statement of task. While the focus of these recommendations is on how to increase the effectiveness of spill response planning and response for spills of diluted bitumen, the committee's recommendations are relevant to other oils that share physical and chemical properties with diluted bitumen (i.e., heavy oils), although non-bitumen heavy oils are beyond the scope of this report.

2

Chemical and Physical Properties of Crude Oils

INTRODUCTION

Crude oil derives, by way of geological processing, from organic material initially buried in sediments at the bottom of ancient lakes and oceans. Crude oil formed at depth in a sedimentary basin migrates upward because of lower density. Many such migrations end with the oil collecting beneath a layer of impermeable rock, also referred to as a "trap," and forming a reservoir that can be tapped by drilling.

If the oil approaches the surface, it cools and comes in contact with groundwater. At the oil-water interface, anaerobic microorganisms degrade the oil in the absence of oxygen. The progressive loss of metabolizable molecules from the oil leads to an increase in viscosity and eventually results in a tarry residue that clogs the pores of the strata through which the oil had been migrating. Over a long duration and with adequate sources of oil from below, enormous deposits of biodegraded oil residue can accumulate. This sequence is how the Alberta oil sands[21] and other oil-sand deposits were formed.

Bitumen is separated from the host rock or sand by heating, which reduces its viscosity so that it can flow to a collection point. Once collected, it is mixed with a diluent so that its viscosity is low enough to allow transport in a transmission pipeline. Such mixtures are called diluted bitumen.

Diluted bitumen are engineered to resemble other crude oils that are transported via pipeline and processed in the same refineries. The composition of diluted bitumen is dependent on several factors, particularly the diluent or diluents chosen and the diluent-to-bitumen ratio. As a result, diluted bitumen has dimensions of variability significantly exceeding those of crude oil from a given source region.[21]

CHEMICAL COMPOSITION OF DILUTED BITUMEN

Diluted bitumen and other crude oils generally contain the same classes of compounds, but the relative abundances of those classes vary widely. Those variations are associated in turn with wide differences in physical and chemical properties. Industry-standard analyses group compounds into four main classes, namely saturated hydrocarbons, aromatic hydrocarbons, resins, and asphaltenes. Saturated hydrocarbons are most abundant in light crude oils, which are the least dense and least viscous. Denser and more viscous crude oils have greater concentrations of other components, including resins and asphaltenes, which contain more polar compounds, often including "heteroatoms" of nitrogen, sulfur, and oxygen as well as carbon and hydrogen.

Even among light or medium crude oils, the relative abundances of specific compounds can vary significantly. The relative abundances will depend on the precise composition of the organic material delivered to the source sediments, the rate and length of time over which the source rock was heated, which inorganic minerals—potential catalysts of specific chemical reactions—were present in the source rock, the distance and details of the migration pathway, and conditions in the reservoir. In Table 2-1 and Figure 2-1, North American crude oils of each type for which data are readily available are provided as representative examples.[22]

From light, to medium and heavy crudes, and on to diluted bitumen, the abundance of saturated hydrocarbons drops 4-fold and the combined abundances of resins and asphaltenes increase 50-fold. These differences

TABLE 2-1 Major Classes of Compounds in Crude Oils, Percentages by Weight

Type of Crude Oil	Saturates	Aromatics	Resins	Asphaltenes
Light Crude[a]	92	8	1	0
Medium Crude[b]	78	15	6	1
Heavy Crude[c]	38	29	20	13
Diluted Bitumen[d]	25	22	33	20

[a]Scotia Light.
[b]West Texas Intermediate.
[c]Sockeye Sour.
[d]Cold Lake Blend.
SOURCE: Hollebone[22]

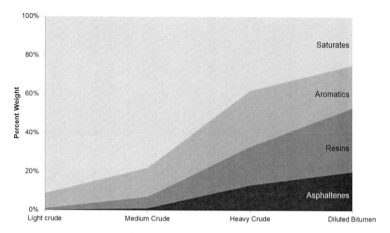

FIGURE 2-1 Components of typical crude oils.

are attributable mainly to the great influence of biodegradation on the heavier crude oil and bitumen.

Saturated Hydrocarbons

Under the anaerobic conditions prevailing during formation of the oil sands, the saturated hydrocarbons are mostly biodegradable, the aromatic hydrocarbons much less so, and the resins and asphaltenes not at all. A heavy crude, or the bitumen from an oil sand, is composed of the residue from a very protracted process whereby microbial action consumes most of the metabolizable saturates.

The saturated hydrocarbon fraction in diluted bitumen thus differs from that in other crude oils because the readily metabolizable molecules are missing. This is seen most dramatically in chemical analyses that reveal the distribution of individual compounds in the crude oil. For example, the graphs in Figure 2-2 show results of parallel analyses of samples of Cold Lake Blend diluted bitumen and Bakken crude oil.[23] The latter is dominated by a strong series of peaks representing its abundant, straight-chain, saturated hydrocarbons. The diluted bitumen, in contrast, is dominated by a hump representing the profusion of branched and cyclic hydrocarbons that are more resistant to biodegradation. These are so numerous and varied that their peaks overlap and they cannot be resolved by this gas chromatographic analysis. The diluted bitumen has a small series of peaks indicating the presence of *some* straight-chain hydrocarbons that derive from the diluent.

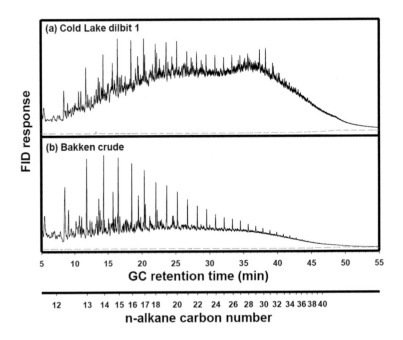

FIGURE 2-2 Gas chromatography (GC) comparison of Cold Lake Blend diluted bitumen and Bakken crude oil, a light crude, from North Dakota (FID, flame ionization detector).
SOURCE: Swarthout, et al.[23]

Aromatic Hydrocarbons

Crude oils contain aromatic hydrocarbons possessing one or more aromatic rings. Those with more than one ring are commonly referred to as polycyclic aromatic hydrocarbons (PAHs). The one-ring compounds are most abundant and are referred to collectively as BTEX, an acronym based on the chemical names of benzene, toluene, ethyl benzene, and xylenes. The most common aromatic hydrocarbons with two rings are naphthalenes. Other commonly measured groups include the three-ring phenanthrenes, dibenzothiophenes, and fluorenes and also the four-ring chrysenes.

The napthalenes and the even larger phenanthrenes are progressively less volatile and soluble compared to BTEX. PAHs are present as unsubstituted or parent forms but the vast majority are alkyl substituted PAHs.

The aromatic hydrocarbons are of interest because of their toxicity. Specific properties and risks are discussed in Chapter 3. In Table 2-2, abundances of commonly measured PAHs in crude oils and in diluted bitumen are listed. Values listed in red mark cases in which the concentration in diluted bitumen exceeds that in the other crude oils indicated.

Resins and Asphaltenes

The resins and asphaltenes characteristic of heavy crudes and diluted bitumen can precipitate from the oil as black sludge and cause numerous problems: clogging well bores, pipelines, and apparatus.[24] Moreover, refining costs increase with the abundances of resins and asphaltenes.[24] For all of those reasons, light and medium crudes have been favored. With increasing pressure on supplies, and with continued improvements in refining processes, heavy crude oils have come into broader use. As shown by Table 2-1, the content of resins and asphaltenes in light and medium crude oils is very much lower than that in heavy crude oils, and lower still than that in diluted bitumen.

The resins and asphaltenes have presented major challenges to chemical analysts.[24] The range of structures and the tendency of the molecules to cluster in larger, multimolecular aggregates make it difficult to determine even rudimentary properties like molecular weight. It has been shown only recently[25] that most individual molecules in the heavy residues have from 30 to 70 carbon atoms. They comprise a complex mixture of polycyclic molecular structures in that range. These molecules tend to stick together, not only in bulk (the property that makes asphalt an attractive paving material) but even at the low concentrations prevailing when samples are injected into analytical instruments. The resulting "nanoaggregates" have masses two to five times higher than those of the molecules of which they are composed. The *apparent* molecular weights are accordingly higher than the *true* molecular weights. Heteroatoms (mainly nitrogen, sulfur, and oxygen) and metals (mainly nickel, vanadium, and iron) are also present in higher relative abundances in the resin and asphaltene fractions than in the saturate fraction. As a result, the heteroatom content of bitumen is higher than that of other crude oils.

Chemical Composition of Diluents

The density and viscosity of raw bitumen are too great to allow transportation by transmission oil pipeline without heating or alteration of the material. To reduce the viscosity and density, a diluent must be added to bitumen to produce an engineered mixture with a density of less than 0.94 g/cm^3 and a viscosity of less than 350 cSt. Additional industry-standard specifications that are largely a function of the operating temperature of the pipeline vary seasonally. Diluents alone do not confer unique chemical or toxicological properties to diluted bitumen;

TABLE 2-2 Concentrations of Parent and Alkylated PAHs, EPA Priority PAHs, and Total Aromatic Compounds in Various Crude Oils

Oil Type	Light	Medium	Heavy	Diluted Bitumen
Oil Sample	**Scotia Light oil[26]**	**West Texas oil[27]**	**Sockeye[27]**	**Cold Lake Blend[28]**
Specific Compounds	**Conc. (µg/g)**	**Conc. (µg/g)**	**Conc. (µg/g)**	**Conc. (µg/g)**
Total EPA 16 Priority PAHs* (µg/g)	139	514	218	176
Total Aromatic Compounds (µg/g)	3,504	7947	5,231	5,384
Parent and Alkylated PAHs				
Sum Naphthalene* C0-4[1]	2692	5172	3422	2099
Sum Phenanthrene* C0-4	351	1295	1078	1242
Sum Dibenzothiophene C0-3	16.3	816	403	1250
Sum Fluorene* C0-3	358	458	184	535
Sum Chrysene* C0-3	17.9	100	60	200
Total parent and alkylated PAHs	3,434	7841	5,147	5,326
Biphenyl	25.9	68.5	34.23	6.58
Acenaphthylene*	3.91	11.08	6.72	2.16
Acenaphthene*	24.2	8.84	7.7	6.93
Anthracene*	1.57	1.00	2.2	N.D.
Fluoranthene*	2.93	2.12	1.22	4.31
Pyrene*	2.55	6.72	5.01	11.3

TABLE 2-2 Continued

Oil Type	Light	Medium	Heavy	Diluted Bitumen
Benz[a] anthracene*	1.41	1.24	3.18	2.52
Benzo[b] fluoranthene*	1.29	1.37	0.98	4.06
Benzo[k] fluoranthene*	0.30	0.37	0.40	0.81
Benzo[e]pyrene	1.33	3.48	1.59	4.10
Benzo[a]pyrene*	0.74	0.25	0.49	3.01
Perylene	1.10	0.12	19.32	7.16
Indeno[1,2,3-cd] pyrene*	0.38	0.18	N.D.	1.89
Dibenzo[ah] anthracene*	0.32	0.18	0.12	0.73
Benzo[ghi] perylene*	1.29	0.50	0.86	2.24

[1]N.D. = not detected. Red values indicate levels in Cold Lake Bitumen that are higher in comparison to Scotia Light and West Texas crude oils. C0 are parent unsubstituted PAHs and C1-C4 are the alkyl PAHs. *Denotes the 16 EPA priority PAHs (naphthalene, phenanthrene, fluorene, and chrysene use the C0 parent levels only).

all crude oils contain similar, light end components. The compositions of diluents, however, can strongly affect the weathering behavior of diluted bitumen, chiefly because the evaporation of a highly volatile diluent will more readily produce a heavy residue.

The individual selection of diluents varies depending on the desired outcome, the current cost of acquiring and transporting the diluent to the bitumen source, and other internal considerations of pipeline operators. Specific information about the diluents used is typically not publicly available. In general, diluents used fall into two broad categories: naturally occurring mixtures of light hydrocarbons, synthetic crude oil, or both.

Synthetic crude oil is produced by upgrading bitumen to reduce its density and viscosity for transport by pipeline. When mixed with bitumen to obtain the required viscosity and density, synthetic crude oils yield a product that can be handled efficiently and economically by conventional heavy oil refineries. A drawback is that supplies of synthetic crudes are limited by the availability of upgraders at the source of extraction and that roughly a 50:50 mixture of bitumen with synthetic crude oil is required to obtain the desired density and viscosity.

The alternative, and more commonly used, diluents are naturally occurring mixtures of light hydrocarbons. These light hydrocarbons are acquired from two sources: ultralight crude oils and gas condensates. Gas condensates are produced by separating most of the C_3 and all of the C_4 and higher hydrocarbons from natural gas. Because the ultralight crudes and gas condensates are less dense and less viscous than synthetic crude oil, diluent-to-bitumen ratios are roughly 30:70. The particular mixture of light hydrocarbons in the diluent can be important in spill response. If the diluent is dominated by lighter compounds (C_4-C_8), it can evaporate more readily in the event of a spill, yielding a dense and viscous residue that must be accounted for in response.

WEATHERING AND ITS EFFECTS ON PHYSICAL PROPERTIES

The behavior of a crude oil or diluted bitumen released into the environment is shaped not only by its chemical composition but also by its physical properties. Those of particular interest are density, viscosity, flash point, and adhesion. Oil spilled into the environment undergoes a series of physical and chemical changes that in combination are termed weathering. Weathering processes occur at different rates, but they begin as soon as oil is spilled and usually proceed most rapidly immediately after the spill. Most weathering processes are highly temperature dependent and will slow to insignificant rates as temperatures approach freezing.

The most important weathering process is evaporation,[29] which accounts for the greatest losses of material. Over a period of several days, a light fuel such as gasoline evaporates completely at temperatures above freezing, whereas only a small percentage of bitumen evaporates. Importantly, properties of the residual oil change as the light components of the oil are removed.

Density

Given that the density of fresh water is 1.00 g/cm^3 at environmental temperatures and the densities of crude oils commonly range from 0.7 to 0.99 g/cm^3 (see Table 2-3), most oils will float on freshwater. Because the density of seawater is 1.03 g/cm^3, even the heaviest oils will usually float on seawater. But evaporative losses of light components can lead to significant increases in density of the residual oil. The densities of some weathered, diluted bitumen and of undiluted bitumen can approach and possibly exceed that of freshwater. Accordingly, those materials can submerge and may sink to the bottom. In this respect, diluted bitumen differs not only from light and medium crude oils, but even from most conventional heavy crude oils. Details are shown in Table 2-3.

TABLE 2-3 Density Comparison of Typical Crude Oils[a]

Type of Crude Oil	Density Before Release	Density After Initial Weathering (mass % loss in weathering)	Density After Additional Weathering (mass % lost in weathering)
Light Crude[b]	0.77	0.80 (25%)	0.84 (64%)
Medium Crude[c]	0.85	0.87 (10%)	0.90 (32%)
Heavy Crude[d]	0.94	0.97 (10%)	0.98 (19%)
Diluted Bitumen[e]	0.92	0.98 (15%)	1.002 (30%)
Bitumen	0.998	1.002 (1%)	1.004 (2%)

[a]Data in g/cm^3 at 15°C; freshwater has a density of 1.00, seawater of 1.03.
[b]Scotia Light.
[c]West Texas Intermediate.
[d]Sockeye Sour.
[e]Cold Lake Blend.
SOURCE: Hollebone[22]

Importantly, as discussed in Chapter 3, as the density of a weathering oil approaches that of water, contact with even small amounts of sand, clay, or other suspended sediment can trigger submergence. For this reason, the density of the oil residue itself (i.e., not including any associated natural particulate matter) does not need to exceed that of water for the residue to sink from the surface.

Viscosity

Viscosity is defined as the resistance to flow of a liquid: the lower the viscosity, the more readily a liquid flows. For example, water has a low viscosity and flows readily, whereas honey, with a high viscosity, flows poorly. The viscosity of oil is largely determined by its content of large, polar molecules, namely resins and asphaltenes. The greater the percentage of light components such as saturates and the lower the amount of asphaltenes, the lower the viscosity. Temperature also affects viscosity, with a lower temperature resulting in a higher viscosity. The variations with temperature are commonly large. Oil that flows readily at 40°C can become a slow-moving, viscous mass at 10°C. Evaporative losses selectively remove lighter components and, consequently, increase the viscosity of the residual oil, as illustrated in Table 2-4.

TABLE 2-4 Viscosity Comparison of Typical Crude Oils[a]

Type of Crude Oil	Viscosity Before Release	Viscosity After Initial Weathering (mass % loss in weathering)	Viscosity After Additional Weathering (mass % lost in weathering)
Light Crude[b]	1	2 (25%)	5 (64%)
Medium Crude[c]	9	16 (10%)	112 (32%)
Heavy Crude[d]	820	8,700 (10%)	475,000 (19%)
Diluted Bitumen[e]	270	6,300 (15%)	50,000 (30%)
Bitumen	260,000	300,000 (1%)	400,000 (2%)

[a]Data in mPa·s.
[b]Scotia Light.
[c]West Texas Intermediate.
[d]Sockeye Sour.
[e]Cold Lake Blend.
SOURCE: Hollebone[22]

Flash Point

The flash point of oil is the temperature at which the liquid produces vapors sufficient for ignition by an open flame. A liquid is considered to be flammable if its flash point is less than 60°C. There is a broad range of flash points for oils and petroleum products, many of which are considered flammable, especially when freshly spilled. Gasoline, which is flammable under all ambient conditions, poses a serious hazard when spilled. Many fresh crude oils and diluted bitumen have an abundance of volatile components and may be flammable for a day or longer after being spilled, depending on the rate at which highly volatile components are lost by evaporation. On the other hand, undiluted bitumen and heavy crude oils typically are not flammable. Table 2-5 provides a quantitative summary of these variations.

Adhesion

The adhesion or "stickiness" of some crude oils has been noted as a problem at several spills. The adhesion of a crude oil to the surfaces of rocks, built surfaces, and vegetation can greatly impede cleanup. Although important in the context of oil spill response, adhesion is a property that is not measured during industry-standard analyses of crude oils. However, a quantitative measure of adhesion has been developed[30] and a comparison of some values appears in Table 2-6. The test measures the mass of oil, or of weathered oil, that will adhere to a steel needle that has been immersed in the sample for 30 min and then allowed to drain for

TABLE 2-5 Flash Point Comparison of Typical Crude Oils[a]

Type of Crude Oil	Flash Point Before Release	Flash Point After Initial Weathering (mass % loss in weathering)	Flash Point After Additional Weathering (mass % lost in weathering)
Light Crude[b]	<−30	23 (25%)	95 (64%)
Medium Crude[c]	−10	33 (10%)	>110 (32%)
Heavy Crude[d]	−3	67 (10%)	>95 (19%)
Diluted Bitumen[e]	<−35	>60 (15%)	>70 (30%)
Bitumen	>100	>100 (1%)	>110 (2%)

[a]Data in °C.
[b]Scotia Light.
[c]West Texas Intermediate.
[d]Sockeye Sour.
[e]Cold Lake Blend.
SOURCE: Hollebone[22]

30 min. As can be seen, diluted bitumen is much more strongly adhesive than light or medium crude oils, or their evaporated residues. The contrast is even greater than it may appear. Not only is the diluted bitumen residue more adhesive, there is much more of it relative to the discharge for a given spill, due to the greater abundances of resins and asphaltenes in diluted bitumen. A comparison of the adhesion for various crude oils is listed in Table 2-6.

TABLE 2-6 Adhesion Comparison of Typical Crude Oils[a]

Type of Crude Oil	Adhesion Before Release	Adhesion After Initial Weathering (mass % loss in weathering)	Adhesion After Additional Weathering (mass % lost in weathering)
Light Crude[b]	0	2 (25%)	9 (64%)
Medium Crude[c]	12	17 (10%)	33 (32%)
Heavy Crude[d]	75	98 (10%)	605 (19%)
Diluted Bitumen[e]	98	146 (6%)	1580 (20%)
Bitumen	575		

[a]Data in g/m².
[b]Scotia Light.
[c]West Texas Intermediate.
[d]Sockeye Sour.
[e]Cold Lake Blend.
SOURCE: Hollebone[22]

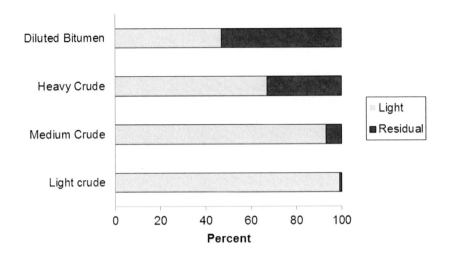

FIGURE 2-3 The relative proportions of light versus residual components in crude oils.
SOURCES: Hollebone[22] and Environment Canada[31]

TABLE 2-7 Comparison of Important Crude Oil Properties

Type of Crude Oil	Adhesion (g/m^2)	Density (g/cm^3)	Viscosity (mPa·s)	Flash point (°C)
Light Crude[a]	0	0.77	1	−30
Weathered[b] Light Crude	9	0.84	5	95
Medium Crude[c]	12	0.85	8	−10
Weathered Medium Crude	33	0.90	112	>110
Heavy Crude[d]	75	0.94	820	−3
Weathered Heavy Crude	600	0.98	475,000	>95
Diluted Bitumen[e]	98	0.92	270	−35
Weathered Diluted Bitumen	1,580	1.002	50,000	>70

[a]Scotia Light.
[b]After additional weathering.
[c]West Texas Intermediate.
[d]Sockeye Sour.
[e]Cold Lake Blend.
SOURCES: Hollebone[22] and Environment Canada[31]

CONCLUSION

Crude oils are mixtures of hydrocarbon compounds ranging from smaller, volatile compounds to very large, nonvolatile compounds. The hydrocarbon structures found in oil include saturates, aromatics, and polar compounds that include resins and asphaltenes. The resins and asphaltenes are largely recalcitrant in the environment. They evaporate, dissolve, and degrade poorly and thus may accumulate as residues after a spill. The percentage of the saturates and aromatics—herein called the light components, in comparison to the heavy, residue-forming resins and asphaltenes—varies with oil type and is summarized in Figure 2-3.

The physical and chemical properties of diluted bitumen differ substantially from those of other crude oils, with key differences highlighted in Table 2-7. The distinct physical and chemical properties of diluted bitumen arise from two components: the bitumen provides the high-molecular-weight components that contribute most to density, viscosity, and adhesion; and the diluent contributes the low-molecular-weight compounds that confer volatility and flammability, and that determine the rate at which evaporation increases the density of the residual oil.

Because diluted bitumen has higher concentrations of resins and asphaltenes than most crude oils, spills of diluted bitumen products will produce relatively larger volumes of persistent residues. Such residues may be produced relatively rapidly when gas condensate has been used as the diluent, and these weathered residues display striking differences in behavior compared to other oils: exceptionally high levels of adhesion, density, and viscosity.[22]

Contrasts between diluted bitumen and other crude oils are strongly enhanced by weathering. Weathered heavy crude and especially weathered diluted bitumen are, for example, much more adhesive than the other oils. The densities of the oils also vary, with weathered heavy oils approaching the density of fresh water and weathered diluted bitumen possibly exceeding the density of fresh water.

3

Environmental Processes, Behavior, and Toxicity of Diluted Bitumen

INTRODUCTION

This chapter is concerned with what happens to crude oil after it is released from a pipeline, and with the potential environmental and ecological consequences of that release. It thus considers the chemical and physical processes affecting the oil and its residues, the resulting behaviors that manifest across various environmental settings, and the toxicity of the spilled and eventual residual oil. The discussion further distinguishes ways in which the behavior of diluted bitumen is similar to or distinct from that of the light and medium crude oils that are commonly transported in U.S. pipelines.

ENVIRONMENTAL PROCESSES

As mentioned in the previous chapter, crude oil released to the environment experiences a host of chemical and physical changes—processes collectively referred to as "weathering." For the purposes of this chapter, weathering processes are divided into three categories based on the nature of the chemical and physical effects on the oil. Chemical processes include photooxidation and biodegradation and cause alteration to molecular structures through the cleavage and formation of covalent bonds, the linkages between atoms. Physical-chemical partitioning processes include evaporation and dissolution. These act without changing molecule structures and partition material, for example, between the atmosphere and a liquid phase. Physical processes include spreading, dispersion, emulsification, adhesion, and sedimentation. All of these change the physical properties and behavior of the oil but do not always partition it between phases or change its molecular structure.

Processes occurring for a crude oil spill on water are outlined in Figure 3-1. The details of each process are discussed in this section. Given the statement of task, the focus is on the processes most relevant to oil spilled in subaerial and aquatic continental and coastal environments, including groundwater, lakes, estuaries, and streams of all sizes. By tracing the flow of the oil, where the origin of the spill represents day 0 of the spill, a general sequence of processes that occur can be summarized.

Chemical Processes

Although many processes act on spilled oil, few processes lead to chemical decomposition. The two processes of greatest relevance to oil spills in the environment are photochemical oxidation and biodegradation. These processes tend to occur slowly over a period of weeks to years and represent the breakdown of oil at the molecular level.

Photooxidation

Photochemical processes result from exposure of spilled oil to sunlight, leading to cleavage and formation of covalent bonds. Oxygen is typically incorporated into the products and thus the term photooxidation is commonly used. These oxidized products include both carbon dioxide and other oxygenated compounds.[32] Typically, aromatic hydrocarbons are transformed more rapidly than alkanes,[33] thereby increasing the relative abundance of resins and asphaltenes in the residual oil.[32,33c] In one set of laboratory experiments, the photooxidation of crude oil in fresh water under direct ultraviolet irradiation showed oxidation of 5% of the branched alkanes, 9% of the linear alkanes, and 37% of the aromatic hydrocarbons.[34]

Although photooxidation can be important when spills receive intense solar exposure, the photooxidation of diluted bitumen in the environment has hardly been studied and its role in the weathering of spilled diluted bitumen is not well understood. Notably, one potential outcome of photooxidation of crude oils is the production of persistent molecules in the environment[32] containing carboxylic acids and alcohols, which may be soluble in water. Thus, these photoproducts may be transported in surface waters or groundwater.[35]

Photochemical enhancement of toxicity has been demonstrated for some polycyclic aromatic hydrocarbons (PAHs) via photomodification and photosensitization. In one study,[36] organisms exposed to PAHs from

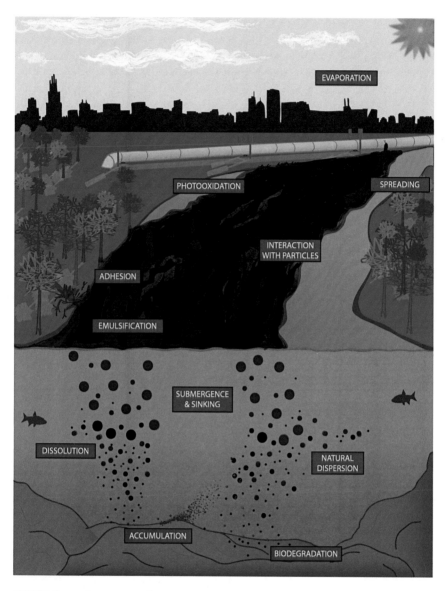

FIGURE 3-1 Processes affecting the composition, amount, and behavior of diluted bitumen.

spilled oil experienced up to a 48-fold greater toxicity when illuminated with natural sunlight instead of standard laboratory light. This mechanism of toxicity is important for early-life-stage and translucent organisms that often accumulate PAHs in their tissues and inhabit surface waters.

Biodegradation

Biodegradation is a process by which living organisms, mainly bacteria, degrade hydrocarbons.[37] Biodegradation can occur either aerobically or anaerobically, with aerobic processes typically occurring more rapidly and extensively. Biodegradation is accelerated in the presence of abundant oxygen and nutrients,[38] moderate temperature and salinity, and reasonable oil-water interfacial surface area.[39] For diluted bitumen, the extent of biodegradation depends on a combination of environmental factors, the proportions of bitumen and diluent, the nature of the diluent, and how fast the diluent is lost to evaporation. Since the deposits from which bitumen is extracted are themselves residues remaining after extensive anaerobic biodegradation, a spill of diluted bitumen may be less susceptible to biodegradation than a comparable spill of light or medium crude oil. However, biodegradation of diluted bitumen in environments containing aerobic bacteria and nutrients warrants further study.[40]

The main classes of crude oil components highlighted in Chapter 2 (see Figure 2-1)—saturates, aromatics, resins, and asphaltenes—provide a useful framework for understanding the relevance of biodegradation to the environmental fate of diluted bitumen. There are no quantitative field studies on the biodegradation of spills of diluted bitumen, but saturates and aromatics are expected to biodegrade within weeks to years.

Conversely, resins and asphaltenes in the bitumen are expected to remain recalcitrant for a longer time.

The U.S. Environmental Protection Agency (USEPA) studied short-term biodegradation in the laboratory using residual oil in sediment from the Kalamazoo River that was collected 19-20 months after a diluted bitumen spill.[41] Over 28 days of aerobic incubation of sediment slurries with inorganic nutrients added, about 25% of the total petroleum hydrocarbons degraded, mostly in the first 14 days. The decreasing rate of biodegradation over the 28-day period suggested that the majority of the spilled oil would not degrade over time scales of at least a few months in spite of the experiment's favorable conditions for bacterial activity.

Physical-Chemical Partitioning Processes

Once crude oil is spilled from a transmission pipeline, its composition can be affected by evaporation of volatile compounds and aqueous dissolution of water-soluble compounds. These processes tend to occur rapidly and strongly impact the composition and behavior of residual spilled oil.

Evaporation

Following a spill that brings oil into contact with the atmosphere, light components will evaporate at relative rates that depend on their volatility. As a result, compounds with greater volatility tend to evaporate from oil more rapidly than those with lesser volatility. This relative relationship tends to hold across various environments but absolute rates vary substantially based on concentrations of the volatile compounds in the oil and on ambient conditions including exposed surface area and volume; temperature of the oil, water, and air; and velocities of the wind current.[42] Fingas[43] has argued that the importance of the wind speed is moderated by the fact that the supply of hydrocarbon molecules to the oil-atmosphere interface is often a limiting step for evaporation.

For diluted bitumen, as with other crude oils, evaporation of light components can occur readily. The relatively light natural-gas condensates often used as the diluent in diluted bitumen are particularly volatile. The loss of volatiles thus leads to a residue strongly resembling the original bitumen, and this is a key behavior that distinguishes diluted bitumen from other commonly transported crude oils. As noted in Chapter 2, such processes increase both density and adhesion of the residual oil. The relationship between density and submergence is considered further here, whereas adhesion is considered in the context of physical processes.

The increase in density that occurs with evaporative loss of the diluent increases the likelihood that the residual oil will submerge beneath the water surface and potentially sink to the bottom.[8,9b] The rate at which density increases will depend on the composition of the diluted bitumen and especially on the nature of the diluent, but significant density increases have been observed to occur over in the first 1-2 weeks of a spill. Diluted bitumen with relatively high proportions of light and heavy hydrocarbons and a paucity of compounds in the C_{15}-C_{25} range, such as Access Western Blend (AWB), Christina Lake, and Borealis Heavy Blend, are expected to achieve a higher density more rapidly with evaporation.

As evidence, King et al.[9b] conducted oil weathering studies in an experimental tank placed outside in Dartmouth, Nova Scotia, Canada (Figure 3-2). As the curves fit to the experimental data in Figure 3-2 show, within 13 days, the density of residues derived from Access Western Blend exceeded that of fresh water, and the oil sank until reaching an underlying layer of saline water. The maximum density of AWB (1.008 g/cm^3) is greater than that of the density of weathered residues of Cold Lake Blend (1.0014 g/cm^3) and it takes less time for AWB to approach its final density

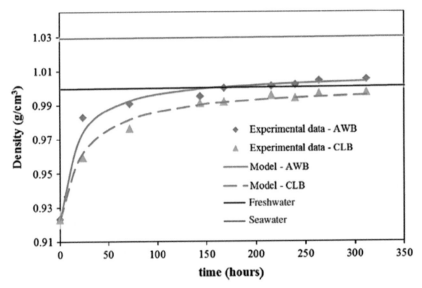

FIGURE 3-2 Observed increases in density for two diluted bitumen (AWB, Access Western Blend; CLB, Cold Lake Blend) added to a flume tank containing seawater, compared to the typical densities of fresh water and seawater. Curved lines show model fits to the observations.
SOURCE: King, et al.[9b]

in comparison with CLB. The time difference could be important when responding to an oil spill, where a difference of 12 hours could result in submergence of the oil.

Dissolution

If the spilled oil is in contact with water, components that are at least slightly soluble in water will be lost by dissolution. For diluted bitumen, the most soluble components originate from the diluent and, because of their volatility, also tend to be readily lost by evaporation. Therefore, unless the oil propagated or spread under water for an extended time, such as in the Deepwater Horizon oil spill,[44] evaporative losses are expected to be larger than dissolution losses. However, evaporative losses may be slower than dissolution if a spill spread under the ice in an ice-covered lake or river. Also, dissolution may become important for diluted bitumen that has percolated to or been released beneath the water table.

Physical Processes

Oil released to the environment may experience changes due to a host of physical processes including spreading, dispersion, emulsification, adhesion, and sedimentation. While these processes do not affect the molecular composition of oil directly, they do result in complex, environment-dependent interactions with the chemical and physical-chemical partitioning processes.

Spreading

On land, spreading of spilled oil is often limited, but when oil reaches a water surface it starts to spread immediately.[45] Unless constrained, the oil will continue to spread out into a thin film, or slick, due mostly to interfacial tension.[45] Spreading rapidly increases the footprint of oil in the environment and can make recovery efforts more difficult, but it also makes the oil more exposed to photooxidation and evaporation.

Small amounts of crude oil can spread into a very thin layer, or "sheen," that is readily visible. Such sheening represents an impairment of water quality that can determine the degree of oil recovery that is required, as happened in the case of the spill in Marshall, Michigan (Box 3-1, Figure 3-3). Submerged or sunken oil residues, mostly out of sight, can still serve as a source of a sheen. The effect can continue for long periods, either spontaneously in response to temperature and water-level changes, or as a result of disturbance of the sediments by animals, boats, etc.

Dispersion

In the context of oil spills, dispersion refers to the entrainment of oil droplets in the water column. The extent of oil dispersion depends on the interfacial tension between oil and water, oil viscosity, and the mixing energy that may be driven by wind, currents, or tides.[46] The interfacial tension between the oil and water does not vary widely among oil types. The range is typically less than twofold. When applied, chemical dispersants can decrease interfacial tension by 10- to 200-fold,[47] allowing a greater proportion of oil to disperse into the water column. The mixing energy varies across environments as well as over space and time in a particular environment (e.g., rivers passing through dams or cataracts, and windy versus calm weather on lakes and coastal marine waters).

The droplet size distribution (DSD) of oil dispersed in water plays an important role in the behavior of oil in the aquatic environment. Larger droplets are more buoyant than smaller droplets and thus rise to the water surface, regardless of whether they were released underwater or released

---

BOX 3-1
Marshall, Michigan: Enbridge

The largest release of diluted bitumen into the environment occurred in July 2010 when the Line 6B pipeline operated by Enbridge Energy Partners LLC ruptured and released an estimated 843,000 gallons of oil into a tributary of the Kalamazoo River near Marshall, Michigan.[12, 52] This spill is recognized as one of the largest, and most costly, inland spills in North America, with estimated costs exceeding $1 billion.

Oil flowed a couple of miles down the tributary and entered the Kalamazoo River, ultimately affecting about 40 miles of stream and river channels. The release occurred when the river was at flood stage and had temporarily inundated its floodplain; shortly thereafter, falling water levels left oil stranded on vegetation and soils up to about a meter above the normal summer river level. The river carries a lower suspended sediment load (i.e., less turbid) than most U.S. rivers and is not particularly turbulent because it has a low elevational gradient.

The U.S. Environmental Protection Agency (USEPA) quickly assumed control of the emergency response under the National Contingency Plan (NCP) and worked with Enbridge and a host of federal, state, and local government agencies.[12] The USEPA remained in this function through 2014, an exceptionally protracted response period. The Michigan Department of Environmental Quality is still engaged in remediation as of 2015.

Initial response focused on capturing and collecting floating oil using conventional techniques such as conventional and sorbent boom, and the majority of the oil was recovered as floating oil or deposits on land, including from the wetland at the source of the release. A major wildlife rehabilitation effort commenced, ultimately cleaning and releasing over 3,000 turtles as well as some mammals and birds. Visible oil on floodplain and riparian vegetation was removed, mostly manually.

Within a few weeks it became apparent that significant amounts of the oil had sunk to the bottom of the river. Recovery of sunken oil increasingly became the focus of response efforts after the initial autumn, although some heavily oiled islands and floodplain areas still required excavation. Oil accumulated on the bot-

---

on the water surface and entrained into the water column by turbulence. In the Deepwater Horizon spill in the Gulf of Mexico, where the oil was released at depth, large droplets (> 1.0 mm) rose almost vertically and reached the surface within hours,[44c,48] while small droplets rose more slowly or became permanently entrained.[49] Studies of surface oil slicks[50] predicted rapid transport of large oil droplets and slow transport of smaller droplets, causing the formation of a comet-shaped oil slick on the water surface. The DSD affects not only the transport but also the fate and toxicity of spilled oil. Increasing the proportion of small droplets results in an increase in the surface area per unit mass of oil, which enhances the tom wherever the river flow slowed down, but particularly behind three areas with manmade dams, and in some oxbows (abandoned river channels).

Detection and quantification of the sunken oil was challenging.[9a] Laboratory measurements of total petroleum hydrocarbons suffered from interferences by natural organic matter. Sunken oil was mapped out using a method called poling, in which the sediments were disturbed with a disk on the end of a pole and the amount of sheen and floating globules that appeared on the surface was estimated. Poling showed that the oil tended to accumulate in areas of slow or no flow and fine, often highly organic sediment. Chemical fingerprinting, requiring an expensive set of measurements, was employed selectively to establish that the sunken oil and the sheen it produced were in fact from the pipeline release and not a legacy from other pollution sources.

With the exception of dredging, proven techniques for recovery of this type of sunken oil in a riverine setting were lacking. Sediment agitation and collection of the resultant sheen was employed extensively in an attempt to recover the sunken oil without sediment removal, but eventually that approach was shown to be inadequate and dredging was conducted in the most extensive contaminated areas. In the later stages of the cleanup, a net environmental benefits analysis was conducted to examine the distribution of sunken oil, as indicated by poling, and to determine whether further recovery attempts were justified against the environmental disturbance of recovery operations.[12] Many small areas of sheen-producing sediments were left alone but continue to be monitored. The USEPA estimates that as much as 80,000 gallons of oil may remain in the sediments, but major recovery efforts ended in 2014.

The toxicity of the spilled oil to fish and wildlife received some study. A histopathological study of fishes conducted shortly after the spill showed evidence for toxic effects, whereas short-term bioassays with invertebrates conducted in later years were less conclusive. Sampling of fishes and benthic invertebrates indicated that communities returned to normal after the first year, although the apparent decline in the first year was clearest for invertebrates, and it is difficult to distinguish the relative roles of direct toxicity of the spilled oil from the effects of cleanup activities (particularly sediment disturbance).

dissolution of hydrocarbons in the water column[44b] and their potential for bioaccumulation via transport across biological membranes.

The time for a slick to break up and reach a stable droplet size distribution (the equilibrium DSD) is also an important factor because the delay enables other processes to act on the oil in a differential way. For example, turbulence could change during this time, leading toward another equilibrium DSD, or particles of a certain size might preferentially interact with mineral or organic particles, thus forming oil-particle aggregates. Figure 3-4 shows how the median droplet size varies with time for oils with various viscosities subjected to agitation at an energy dissipation

FIGURE 3-3 Sheening in the Kalamazoo River, the site of a diluted bitumen spill (see Box 3-1). Left: Sheen emanating from an island after a modest rise in river level (the spill occurred at a higher river level). Right: Sheen generated from sunken oil upon disturbance of the sediments. Photo credit: USEPA.

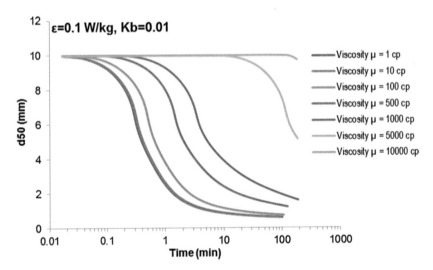

FIGURE 3-4 Evolution of the median droplet diameter (d_{50}) with time as a function of the oil viscosity, based on model simulations.[53] High-viscosity oils would require hours to disperse into an equilibrium droplet size distribution.

rate (ε, W/kg) representative of spilling breaking waves. Even under these conditions, which can be expected to be rare in inland waters, but more likely in a coastal environment, oils with viscosities >10,000 cP require many hours to approach equilibrium, leaving ample opportunity for other processes to act.

Emulsification

Emulsification is the process by which one liquid is dispersed into another one in the form of small droplets.[51] To be called an emulsion, the product must have some physical stability; otherwise the process is called water uptake rather than emulsification. Water droplets can remain in an emulsified oil layer in a stable form and the resulting material, sometimes called mousse or chocolate mousse because of its appearance, has properties differing strongly from those of the parent oil.

Water can be present in oil in five distinct ways. First, some oils contain about 1% water as soluble water. This water does not significantly change the physical or chemical properties of the oil. Second, oil can contain water droplets without forming a stable emulsion. They are formed when water droplets are incorporated into oil by wave action and the mixture is not viscous enough to prevent droplets from separating from the oil.

Mesostable emulsions represent the third way that water can be present in oil and are formed when the small droplets of water are stabilized by a combination of the viscosity of the oil and the interfacial action of asphaltenes and resins. The viscosities of mesostable emulsions are 20-80 times higher than that of the starting oil. These emulsions generally break down within a few days into oil and water or sometimes into water, oil, and emulsion remnants.[51] Mesostable emulsions are viscous liquids that are reddish brown in color.

The fourth way that water exists in oil is in the form of stable emulsions, which form in a way similar to mesostable emulsions and persist because the concentration of asphaltenes and resins is high enough to stabilize the oil-water interface. The viscosity of stable emulsions is 800-1,000 times higher than that of the starting oil. This emulsion will remain stable for weeks and even months after formation. Stable emulsions are reddish brown in color and appear to be nearly solid. Because of their high viscosity and near solidity, these emulsions do not spread and tend to remain in lumps or mats on water or shore.

The fifth way that oil can contain water is by viscosity entrainment. If the viscosity of the oil is such that droplets can penetrate, but will only slowly migrate downward, the oil can contain about 30% to 40% water as long as it is in a turbulent water body. Once the water calms or the oil is removed, the entrained water slowly drains. Typically most of the water will be gone before about 2 days.

The formation of emulsions is an important event in an oil spill. First, and most importantly, emulsification substantially increases the actual volume of the spill. Emulsions of all types initially contain about 50% to 70% water and thus, when emulsions are formed, the volume of the spill can be more than tripled. Even more significantly, the viscosity of the oil increases by as much as 1,000 times, depending on the type of emulsion formed. For example, oil that has the viscosity of a motor oil can triple in volume and become almost solid through the process of emulsification. These increases in volume and viscosity make cleanup operations more difficult. Oil in stable emulsions is difficult or impossible to disperse, to recover with skimmers, or to burn. Mesostable emulsions are relatively easy to break down, whereas stable emulsions may take months or years to break down naturally. Emulsion formation also changes the fate of the oil. When oil forms stable or mesostable emulsions, evaporation slows considerably. Biodegradation also slows. The dissolution of soluble components from oil may also cease once emulsification has occurred.

Adhesion

Adhesion was considered in Chapter 2 as a physical property of oil, and diluted bitumen tends to be more adhesive than other commonly transported crude oils. In this section, adhesion is further considered in the context of weathering, as an environmental process that changes the physical properties and behavior of oil.

When diluted bitumen is spilled and the diluent evaporates, the residual oil will increasingly adhere to surfaces. Many inland and coastal waters contain submerged, floating, and emergent aquatic vegetation and debris that would provide surfaces for adhesion of bitumen. In addition, various animals may also become coated with oil, including turtles, amphibians, insects, and mammalian species (see Figure 3-5). Residual bitumen can be difficult to remove from tree trunks and other biota, rocks, concrete, and manufactured surfaces (Figure 3-5). Strongly adhesive behavior is not unique to bitumen, as some heavy oils with high contents of resins and asphaltenes can also become highly adhesive following weathering. However, diluted bitumen is unique in terms of the rate at which its physical properties can begin to revert back to those of bitumen, due to the evaporative loss of volatile hydrocarbons that comprise the diluent. The emergence of strong adhesion following the evaporative loss of volatile components can impede recovery efforts and, as discussed further below, is expected to increase the tendency of the residue to adhere to particulate matter and sink.

Sedimentation

In the context of this report, sedimentation refers to oil that sinks and comes to rest on the underlying bed in an aquatic setting. Sedimentation may occur by several routes, including an increase in density of the oil through physical-chemical partitioning or chemical processes, the adhe-

(a)

(b)

(c)

(d)

FIGURE 3-5 Biota coated in adhesive oil: (a) Oil that coated the shoreline after the Marshall spill into the Kalamazoo River in Michigan (Box 3-1) as the river level fell in August 2010. Photo credit: S. Hamilton. (b) Residual bitumen oil on a tree along the Kalamazoo River in September 2015, over 5 years after the spill. Photo credit: S. Hamilton. (c) Oiled map turtle from the Kalamazoo River, collected shortly after the spill. Photo credit: Michigan Department of Natural Resources. (d) Ladybird beetle collected at a natural hydrocarbon seep off Coal Oil Point, Santa Barbara, CA. Like bitumen, the seep oil is highly adhesive because of its high resin and asphaltene content. Photo credit: D. Valentine and R. Harwood.

sion of entrained (dispersed) droplets of oil to the bed, and formation of oil-particle aggregates (OPAs) of sufficient density to submerge. This section focuses on OPAs, although the high density of weathered diluted bitumen is expected to increase sedimentation by all three routes, relative to commonly transported crude oils.

Aggregation of oil with natural particulate matter can cause submergence of an initially floating oil.[9a] There are two major types of OPAs: oil droplets coated by small particles[54] and oil trapped within or adhering to large particles (Figure 3-6). The first type is more common and has

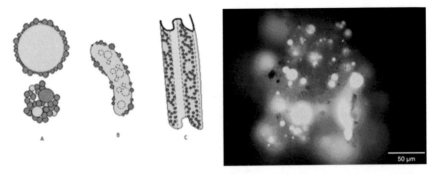

FIGURE 3-6 Left: Types of oil-particle aggregates (OPAs): A, single and multiple oil droplets (yellow) aggregated with natural particles (blue); B, solid aggregate of a large, usually elongated mass of oil with interior particles (dashed blue circles); and C, flake aggregate of thin layers of clay that incorporate oil and fold up (c.f. Stoffyn-Egli and Lee[55] and Fitzpatrick, et al[56]).
Right: Kalamazoo River sediment spiked with weathered source oil from the Marhsall, MI spill after 48 hours, under ultraviolet epifluorescence microscopy at 320× magnification.[57] Images from Fitzpatrick et al.[9a]

been studied in some detail. The particles on the oil droplet surfaces prevent coalescence with other droplets, thus stabilizing the suspension. The OPAs tend to separate from the oil droplets from which they form, and then they may sink because the aggregate of residual oil (density near, or slightly less than, that of water) with inorganic particles (density at least twice that of water) yields an aggregate particle density that is neutrally or negatively buoyant in water.

The formation of aggregates depends on the viscosity of the oil droplet, the surface areas and mineralogy of the particles, and the salinity of the water.[58] Salinity enhances the formation of OPAs, but the onset of aggregation is very steep, becoming important at salinities as low as 1/200 that of seawater.[54b, 58b] The extent of adsorption also depends on the surface properties of the sediment particles and other particles such as algal cells. The relative *quantity* of residual oil produced by weathering is from 5- to 50-fold greater than the quantity produced by weathering of equal volumes of light and medium crude oils.

In quiescent water, the density of the OPA determines whether it remains at the surface or sinks to the bottom. However, when turbulence is present, the suspension of the OPA in the water is due to the balance between hydrodynamic forces that tend to lift the OPA and keep it in suspension versus the gravitational force on the OPA. This means that density alone does not determine the location of the OPA; its size and shape are also important. A large OPA would be less prone to displacement by turbulence than a smaller one and, thus, tends to settle faster than a smaller OPA. Because viscous oils tend to break into large droplets,[59] the OPAs resulting from these oils would tend to be large, and if their density is greater than that of the receiving water, they would settle to the bottom. This formation of large OPAs is a major distinction between diluted bitumen, which rapidly weathers to a viscous residue, and commonly transported crude oils.

Two recent laboratory studies dealt specifically with the formation and sedimentation of oil-particle aggregates in fresh water. The results of the first of these studies[60] are fully consistent with the second, far more detailed report, by Waterman et al.,[61] which focused specifically on the mechanisms underlying the submergence and deposition of diluted bitumen residues in the bed of the Kalamazoo River. Cold Lake Blend, a diluted bitumen, was added to pure water and agitated by bubbling air at 21°C. The study found that it lost at most 17.4% of its mass and reached a density of 0.993 g/cm^3, thus remaining buoyant.[61] However, when the diluted bitumen was added to a mixture of Kalamazoo River water and sediment collected upstream from the spill site, the results were very different. Abundant OPAs formed quickly, on time scales of minutes to an hour, and included both inorganic and organic particles. Measured settling velocities ranged from 1 to 11 mm/s, with most being around 2 mm/s. This observation—that aggregates formed readily in a natural, fresh water—is not in conflict with laboratory studies showing that aggregates form slowly, if at all, in nonsaline water. The natural water is definitely "fresh," but its content of ions derived from clays and soils is high enough to allow the formation of aggregates. Specifically, the electrical conductance of the water from the Kalamazoo River was 640 µS/cm, typical of North American river waters and indicating a salinity about 1/100 that of seawater.

Water temperature and salinity are important determinants of the propensity of residual bitumen to submerge. Short[62] reviewed existing experimental studies of the submergence of weathered bitumen and concluded that the studies had not sufficiently considered the differential effects of temperature on the densities of bitumen products versus that of freshwater or seawater. The density of bitumen increases faster with decreasing temperature than does the density of water, and therefore a weathered bitumen may sink in cooler water while floating in warmer water, yet the studies did not investigate a range of temperatures. Furthermore, Short[62] noted that where there is salinity stratification with fresh or brackish water overlying seawater, as is particularly common at freshwater inlets to coastal marine zones, submerged oil may accumulate at density interfaces beneath the surface. Such interfaces can be dynamic under the influence of winds, tides, and changing freshwater discharge, presenting a particularly challenging situation for crude oil detection and recovery.

In sum, in a diluted bitumen spill subject to weathering, there is much more residue and its density is much closer to that of water, a combination that is likely to translate to enhanced OPA formation and oil submergence relative to commonly transported crude oils.

ENVIRONMENTAL BEHAVIOR

This section addresses how oil spills behave in and potentially impact specific environments, with particular attention to how spills of diluted bitumen may be similar to or distinct from those of other commonly transported crude oils. The toxicity of diluted bitumen is reviewed in a following section.

Water bodies merit particular attention. Pipelines traverse innumerable inland water bodies ranging from headwater streams to wetlands to larger rivers and lakes. Spills can enter water directly or indirectly via overland flow or transport in groundwater. These inland waters are immensely diverse. Once an oil spill contacts water bodies, it may be more difficult to recover, it can potentially be transported quickly from the site of release, it presents a hazard to aquatic life (see toxicity discussion below), and it may impair human uses of the water body.

The land-water interface often presents special challenges for oil spill cleanup. Water levels in lakes, rivers, and coastal zones vary under the influence of precipitation and runoff, dam operations, and/or tides. Accordingly, oil floating on the water can subsequently become stranded in intermittently flooded areas, from which it could be remobilized later. The land-water interface is often disproportionately important to its area, serving as habitat for fish and wildlife and providing recreational and aesthetic values for people.

Spills on Land and into Groundwater

When diluted bitumen or other crude oil is released onto land, or into soils beneath the ground surface, its movement in the ground depends on the soil type, oil viscosity, and the depth to the water table. Oil percolates deeper in coarse sediments, such as gravel, than in fine sediments, such as fine sand, silt, and clay. Extensive studies of a spill of crude oil at Bemidji, Minnesota (Box 3-2), have provided an in-depth understanding of the behavior of crude oil in a subsurface setting. If crude oil is sufficiently fluid to move in the subsurface, it would migrate downward until reaching the water table, where it will sit above the water and could

BOX 3-2
Groundwater Contamination in Bemidji, Minnesota

One of the most comprehensively studied sites for crude oil transport and fate in groundwater has been at Bemidji, Minnesota, where 10,700 barrels of crude oil, characterized as light to medium,[67] spilled from a failed weld on a pipeline in August 1979. About 2,500 barrels seeped into the subsurface and could not be recovered. Remediation of this site remains a challenge today. In 2011, reviewing three decades of remediation and investigation, scientists from the U.S. Geological Survey wrote, "Considerable volumes of NAPL oil [non-aqueous-phase liquid oil] still remain in the subsurface despite 30 years of volatilization, dissolution, and biodegradation, and 5 years of pump-and-skim remediation."[68] The current approach for the contaminated Bemidji aquifer is to rely on natural attenuation, including both aerobic and anaerobic biodegradation by microorganisms.[64]

Some of the liquid oil reached the groundwater table, where it has floated on top of the water, while some was retained by sediments in the unsaturated zone. As depicted in the cross-section figure below, various processes have affected the amount and composition of the subsurface oil, including differential sorption to the soils and underlying glacial deposits, evaporation of volatile components into the soil air space, dissolution of the more soluble components into groundwater, and biodegradation. Natural attenuation by biodegradation has substantially reduced the amount of oil and slowed its movement relative to the underlying groundwater, even though rates are limited by the availability of oxygen and other oxidants.

The oil has gradually migrated in the direction of groundwater flow, yielding a plume of liquid oil, an associated vapor plume in the overlying unsaturated zone, and a plume of dissolved constituents that moves with the groundwater. The different zones in the figure below have distinct concentrations of petroleum contaminants, dissolved oxygen, and the products of microbial reactions that degrade the petroleum compounds. Research at the Bemidji site has provided a wealth of information on the rates and limitations of the microbial processes that biodegrade oil, and this information has been integrated into a mathematical model that informs decisions about remediation of subsurface oil spills.

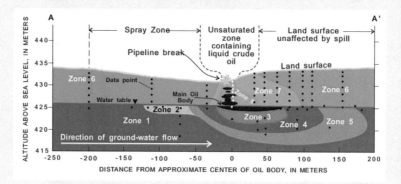

SOURCE: Delin et al.[69]

propagate horizontally along the groundwater flow gradient.[63] In most instances, the underground migration of the free phase oil (or diluted bitumen) is limited, and many small crude oil spills onto land have been remediated by excavating the contaminated soils. In the case of diluted bitumen, if spilled oil is exposed to the atmosphere, the evaporation of the diluent and the light fractions increases the oil viscosity and thus limits oil migration in the ground. However, if not lost to evaporation, the light components of crude oil, especially those that are more soluble in water such as benzene, toluene, ethyl benzene, xylenes (BTEX), and naphthalenes, could migrate relatively easily in the aquifer, especially if the aquifer material contains only small amounts of organic matter. Examples of such materials include gravel and sand aquifers, typical of glacial outwash. For organic-rich soils, water-soluble oil components could be adsorbed onto sediments, depending on their water-to-octanol (K_{OW}) partition coefficients.[63] A concern with crude oil spills underground is the mobilization of trace metals such as arsenic, as has been observed at the Bemidji site in Minnesota.[64]

If oil were spilled into karst aquifers, where transmission of groundwater can be exceptionally rapid, its behavior might be expected to be comparable to that in surface waters. Rapid migration of the oil could pose a threat to water supplies for drinking or irrigation. However, to date there is no experience with diluted bitumen spills into karst aquifers.

The surface soils at the site of the release of diluted bitumen from a pipeline near Marshall, MI, as well as the downstream floodplains of the Kalamazoo River, were significantly contaminated (Box 3-1). At the release site, diluted bitumen saturated soils in a small wetland before overflowing into a creek. Large amounts of residual oil were deposited on downstream floodplains as river levels fell. Much adhered to vegetation and other surfaces as the diluent evaporated. Residual oil did not move long distances in groundwater, presumably because the groundwater flows to the river from the surroundings in almost all of the affected reaches.[65] Had the flows been from the river to the alluvium, a common situation, groundwater contamination may have been more pervasive.

The ecological impacts of the oil deposited on land at the Marshall, MI spill were not studied in depth. Because remobilization of oil by subsequent floods was a major concern, oil-coated vegetation and soils were largely removed within months wherever oil deposits were visible. Little immediate mortality of wildlife was reported. Removal of oily vegetation and soils severely disturbed the riparian and floodplain ecosystems, even though this was done manually in all but the most heavily oiled locations, where excavation was necessary.

In terrestrial ecosystems, spills of diluted bitumen and of light or medium crude oils present similar challenges for cleanup. Contaminated soils, vegetation, and man-made surfaces must be removed or cleaned. The most distinctive feature of diluted bitumen in this regard is its tendency to be strongly adhesive and to coat surfaces, and the difficulty of washing it off. Contact with the sticky bitumen by wildlife and people would increase exposure although toxicity implications are not clear given limited data (see Toxicity section, Chapter 3). The bitumen would be expected to be highly persistent due to its abundance and resistance to dissolution or biodegradation.

Inland Lakes and Reservoirs

Inland lakes and reservoirs vary greatly in depth, exposure to wind and currents, suspended matter, and submerged and emergent aquatic vegetation, all of which potentially affect the behavior and impacts of oil spilled into these environments. Spilled oil may enter a lake or reservoir along a shoreline, or from tributaries. Submerged oil carried in suspension (including as OPAs) by streams and rivers may accumulate in depositional areas once the water enters a lake or reservoir, and this is particularly pertinent to spills of diluted bitumen because of their greater propensity to form OPAs, as discussed above. Lakes and reservoirs are often used as water supplies and for recreation, thus raising the possibility of risks to the public, and they provide important habitat for fish and wildlife.

The 2010 spill in Marshall, MI (Box 3-1) reached a reservoir known as Morrow Lake, and that location provides our only experience to date with a diluted bitumen spill in a reservoir. The extensive delta at the upper end of the reservoir, which begins 60 km downstream of the pipeline release, accumulated submerged oil. Sheening was observed during warm months for 3 years after the spill. A hydrodynamic model indicated the potential for resuspension of the sediment containing oil and its movement further into the reservoir.[66] The most heavily oiled sediments had to be dredged in 2014.

The Great Lakes system of the U.S. and Canada has distinct characteristics that would affect the behavior and impacts of an oil spill. Transmission pipelines capable of transporting diluted bitumen products cross the Great Lakes system at two points: the Straits of Mackinac between Lake Michigan and Lake Huron,[70] and the St. Clair River upstream of Detroit and Lake Erie. A release at either the Mackinac Straits or the St. Clair River would lead to movement of oil into the lakes. Additionally, pipelines cross many streams and rivers that flow short distances to either the southwestern shores of Lake Superior or the southern shores of Lake Michigan.

Currents can be complex in the Great Lakes, with currents in the Straits of Mackinac depending on relative water levels of Lakes Michigan and Huron as well as on wind speed and direction. It could be very difficult to anticipate the movement of the spilled oil and to recover the oil, even at the surface, due to the expansive area and potential for strong wave action. Ice cover during winter could impede detection and recovery of spilled oil.

Wetlands

Pipelines often traverse wetlands, which are common landscapes in the U.S. in all but the arid West. Wetlands are highly variable ecosystems in terms of their hydrology (water sources, flow, and connectivity with water bodies) and ecology (vegetation and animal life). An oil release into a wetland may affect a limited area if the system is hydrologically isolated or has only seasonal flow. However, wetlands commonly occur along land-water interfaces of lakes, rivers, and coastal marine waters, and spilled oil could spread from wetlands into adjacent waters, or vice versa, if they are connected via surface water flow.

Wetlands are valued and protected because they provide important ecosystem services and can harbor rare or threatened plants and animals. Oil spills into wetlands present challenges because of the sensitivity of these ecosystems;[71] for example, in Marshall, MI (Box 3-1), oil trapped in floodplain swamps and marshes had to be recovered manually by cutting vegetation and scraping soils wherever possible, and care had to be taken to not alter the natural hydrology (e.g., by building access roads).

Unless the water is deep enough that submergence of oil poses an issue, diluted bitumen and commonly transported crude oils pose many similar challenges following release in wetlands, but three factors specific to diluted bitumen can be mentioned. First, the amount of residual oil will be large compared to those produced by spills of conventional crude oils. Second, the increased level of adhesion of bitumen may complicate operations in a wetland environment. And third, diluted bitumen residues may persist longer in wetlands, as in other environments, because of their resistance to biodegradation.

Streams and Rivers

Streams and rivers vary in several ways that are important for the transport and fate of oil released into them. Key factors include gradients and velocities of flow; the type and concentration of suspended particulate matter; the types and abundance of underwater algae and plants; the extents and types of zones in which oil may be deposited, such as sedimentation in impoundments and side channels; and variations in flow and water levels. Submerged oil carried by streams and rivers can continue to move downstream until reaching depositional zones in water bodies or floodplains, where it can settle in relatively extensive areas.

The transport of oil in rivers could be also affected by whether the reach is gaining or losing water by exchange with groundwater. Rivers may transport sediment (and associated crude oil) by wash load, suspended load, and bed load.[72] Wash load consists of very fine particles that are relatively evenly distributed through the water column; examples include suspended clays and organic matter. The wash load may represent those particles that would interact with spilled diluted bitumen and influence its initial behavior. Knowledge of bed material composition does not allow one to predict wash load transport. Suspended load is the load that is suspended in the water column but still interacts with the bed. The interaction usually occurs at riffles causing the shear stress to increase and sediments to be suspended from the lee side of the ripples. Bed load has continuous contact with the bed and has a direct relation to the turbulence along the bottom in rivers. While both suspended and bed loads are commonly predicted using sediment transport models, the wash load is not predicted by these models, limiting the prediction of OPA formation in rivers. Furthermore, prediction requires a better understanding of which types of sediment are most likely aggregate with oil to form OPAs, and how those OPAs are likely to be transported in streams and rivers. The answers to these questions are likely to be different for diluted bitumen compared to other commonly transported crude oils.

Estuaries and Coastal Zones

Pipeline spills can readily reach coastal zones if they occur in tributaries or along estuaries, or at coastal processing facilities. Coastal zones in the U.S. where diluted bitumen spills could occur in the future as diluted bitumen pipelines and infrastructure expand include Anacortes in the Pacific Northwest, the Gulf Coast, and Portland, Maine. Each of these regions has distinct physical and ecological characteristics that would influence the fate and ecological effects of spilled diluted bitumen. Many oil spills have occurred in coastal areas[29] and have included heavy refined oil products, though until recently the only one in North America that involved diluted bitumen occurred in Burnaby, British Columbia. The recent Refugio oil spill in California also entered a coastal area and all active cleanup for this spill ended August 31, 2015 (Box 3-3).

Estuaries occur at the interface of terrestrial and marine systems. Estuaries are characterized by periodic reversals of flow due to tides and

Box 3-3
Santa Barbara, California: Refugio Pipeline Spill

On May 19, 2015, a pipeline rupture occurred near Santa Barbara, California, on Line 901, operated by Plains All American Pipeline L.P. The heavy crude oil was discharged from a heated transmission pipeline and comprised a blend of oils from four nearby platforms. Despite the chemical and physical differences between the heavy crude oil spilled at Refugio and diluted bitumen, several aspects of the event provide for a valuable point of comparison for this study, which led to a site visit from members of the study committee.

Unanticipated Complexities

The May 19, 2015, rupture to Line 901 occurred on the inland side of a four-lane highway during daylight hours, though an earthen berm paralleling the highway prevented motorists from seeing the oil. The discharged oil pooled in a local depression and then overflowed to a drainage culvert that ran under the berm. The oil flowed unseen to a second drainage culvert that ran underneath the freeway, to a drainage that led under a railway line, down a gulley to a cliff, and then to a sand and gravel beach from which it entered the ocean. The total discharge was estimated at 101,000 gallons, with 20,000 gallons estimated to have reached the ocean.

The partitioning of the discharge between land and ocean dictated response actions from both USEPA and the U.S. Coast Guard, and a joint command structure was established in which two federal On-Scene Coordinators, one from each agency, shared the lead responsibilities for the response. The bifurcation of discharge and the resulting structure of the Unified Command underscore the importance of response planning for diluted bitumen spills in a terrestrial setting that might subsequently affect lacustrine, riverine, and/or coastal marine environments. This incident further underscores the need for coordinated planning among federal agencies for response to diluted bitumen spills.

by the presence of freshwater-saltwater mixing zones. Large estuaries can be strongly affected by wind and waves. The interaction of fresh and saline waters in estuaries is important not only for physical transport of spilled crude oil but also because mixing of these waters often results in the flocculation and settling of clay particles. Where crude oil is associated with these particles, salinity gradients can be important in controlling the sinking rate.

Coastal zones are characterized by three-dimensional flows due to waves, longshore drift, and rip currents. Although there are numerous regional ocean models, their highest spatial resolution tends to be around

Coastal Zone Transport and Submerged Oil

The product spilled from Line 901 was a heavy crude oil and much of the mass floated at the sea surface. The heavy crude oil that reached the ocean was initially transported along the coast to the east, driven by longshore drift, and opposite to the direction of the prevailing offshore current. This transport gave way to heavy oiling of beaches extending several miles to the east of the rupture. Transport offshore by the prevailing current and then to the southeast caused oiling events and beach closures in Ventura and Los Angeles Counties, up to 100 miles away. Scientists identified submerged and sunken oil in near-shore reefs and kelp beds, near the most heavily oiled beaches. The submerged oil was presumably ballasted by mineral and organic material to which it had adhered while on the beach and in the surf zone. The documentation of submerged oil by scientists triggered response officials to conduct a formal assessment of submerged oil, the result of which was a finding that quantities were too low to warrant removal efforts.

Photo credit: D. Valentine

1.0 km, a scale that is too large to quantify the processes occurring in the coastal zone. While waves in coastal areas tend to bring the crude oil to the beach, there are situations where regional, longshore, and rip currents transport crude oil offshore and parallel to the shoreline and deposit it miles away from where it was released (Box 3-3).

One distinct characteristic of coastal waters (including estuaries) that affects the potential for crude oil to submerge is salinity, which ranges from fresh to oceanic levels and can fluctuate seasonally or daily due to tides and river discharge. Salinity affects the density of the water, and as noted earlier, OPAs that may sink in brackish water can remain in suspension in more saline water. Water flow in coastal zones is often complex and this extends to spatial and temporal patterns of salinity. Tides cause bidirectional flow of water in channelized areas and coastal currents are also commonly induced by waves. Wave action can be produced by local winds and also ocean swells. Freshwater inputs may lie above saltwater. Suspended particulate matter, composed of mixtures of inorganic sediment, detrital organic particles, and algae, also is spatially and temporally variable in coastal waters but often occurs at high concentrations compared to offshore waters. All of these features complicate the prediction of the behavior of an oil spill in these environments, including its propensity to form OPAs and become submerged.

There has been only one significant release of diluted bitumen into a coastal marine zone—the Burnaby Inlet in British Columbia (Box 3-4)—and in that case most of the spilled oil was recovered quickly. However, there is a great deal of experience with crude oil spills in coastal zones because of the high tanker, barge, pipeline, and refinery activities in close proximity. These have included spills where the crude oil became submerged or sank mostly as a result of oil-particle interactions.[46a]

Beaches

Crude oil penetration and persistence in beaches (or river banks and bars) is affected by both crude oil properties and beach hydrodynamics. The mobility of crude oil in porous media such as sand is inversely proportional to its viscosity and, thus, high-viscosity oils (such as weathered diluted bitumen) tend to penetrate very little within the sediment. High adhesion of heavy crude and weathered diluted bitumen oils further reduces their mobility in porous media, making them as amenable to recovery as other crude oils.

In tidally influenced beaches, crude oil deposition is usually highest in the upper intertidal zone and decreases moving toward the lower intertidal zone. However, crude oil buried in the upper intertidal zone tends to weather faster than that in the lower intertidal zone due to high water exchange resulting from tidal action and often from local groundwater inputs that replenish oxygen and nutrients.[73] For example, groundwater movement was found to be a major factor in the persistence of the *Exxon Valdez* oil in Alaska beaches.[38b]

BOX 3-4
Burnaby, British Columbia: Kinder Morgan

On July 24, 2007, approximately 1,400 barrels (58,800 gallons) of Albian heavy synthetic crude oil leaked from the Westridge Transfer Line in Burnaby, British Columbia. The spill resulted from an excavator bucket striking the pipeline during excavation for a new storm sewer line. The pipeline was operated by Kinder Morgan Canada and owned by Trans Mountain Pipeline L.P. The pipeline linked the Burnaby terminal to the Westridge Dock, where oil could be loaded to tankers.[37b] After the oil was spilled, it flowed in Burnaby's storm sewer systems until it reached Burrard Inlet.[37b] In total, 11 houses were sprayed by the rupture, 50 properties were affected, 250 residents voluntarily left, and the Burrard Inlet's marine environment and 1,200 m of shoreline were affected by the spill.[37b]

Cleanup took months and cost roughly $15 million and resulted in the recovery of approximately 1,321 barrels of oil.[75] Responders used three distinct methods to recover the oil, based on the oil's location. In residential areas, peat moss was used successfully to absorb oil on land. In storm sewers, oil in the storm sewers was vacuumed up. Much of the oil was collected in the pump station. And finally, in Burrard Inlet, booming was used to contain oil around the release points, skimmers and absorbent pads were used to remove oil, and tarballs, debris, and oiled macroalgae were manually removed.

To treat the oil that had adhered to the shoreline, responders successfully used the chemical shoreline cleaner Corexit 9580.[76] An estimated 35 barrels of oil were not recovered and were considered to be released to the marine environment.

The recovery effort during the Burnaby Harbor spill was relatively successful. Because the synthetic crude traveled on a predictable path through the storm sewer system, the responders were able to set up booms in a quick and efficient manner. The effects of the spill were limited due to favorable conditions for recovery:[77]

- There was sunny weather with little stormwater flow (slowed the movement of oil in storm drains and facilitated evaporation of oil).
- The spill occurred on slack tide (incoming tide helped keep the oil near shore while booms were placed, and helped limit the movement of oil in the Inlet).
- It occurred outside the primary migration and overwintering period for birds, and after the breeding bird season.
- It occurred prior to the main period of salmon migration to creeks and rivers for spawning.

There were no reports of the oil sinking or becoming submerged in the water column.

TOXICITY

Toxicity of Commonly Transported Crude Oils

As described in Chapter 2 and displayed in Table 2-1, crude oils contain a variety of chemical constituents whose percentage compositions differ widely depending upon the specific product or blend, and thus generalizations about toxicity are difficult. However, the toxicity of typical constituents of crude oil has been demonstrated by many studies. Specific toxic compounds include the monoaromatic hydrocarbons comprising BTEX. The acute toxicity, respiratory issues, and potential carcinogenic properties of BTEX are of concern regarding the health of humans and of wildlife. Various diluted bitumen products show lower BTEX levels (percentage volume)[74] compared to light and medium crudes, with values similar to those of heavy crudes (Table 3-1).

Other crude oil components of interest include the low-molecular-weight linear alkanes, other monoaromatics, and 2-ring PAHs that are often of acute aquatic toxicity concern, mainly due to acute narcosis-based mechanisms of toxicity. Also of concern are the 3- to 5-ring (unsubstituted

TABLE 3-1 Five-Year Average Concentrations and Ranges of BTEX in Various Crude Oils in Percent Volume[ii]

	Light Crude	Medium Crude	Heavy Crude	Diluted Bitumen
Benzene	0.22 (<0.01-0.38)	0.42 (0.14-0.77)	0.12 (0.02-0.21)	0.16 (0.06-0.28)
Ethylbenzene	0.26 (0.24-0.27)	0.35 (0.19-0.60)	0.11 (0.05-0.18)	0.07 (0.04-0.11)
Toluene	0.67 (0.03-1.14)	0.87 (0.29-1.34)	0.26 (0.12-0.45)	0.32 (0.18-0.47)
Xylenes	0.99 (0.18-1.46)	0.77 (0.43-1.09)	0.36 (0.23-0.49)	0.33 ± 0.05 (0.27-0.43)
TOTAL BTEX	2.10 (0.24-3.26)	2.34 (0.97-3.11)	0.84 (0.56-1.26)	0.89 (0.64-1.16)

[ii]Data are from the Crudemonitor database accessed May 2015. Values in parentheses list the range of concentrations across oils in each category. All crude oils with data spanning at least two years were included for comparison. The crude oils include five light crude oils (BC light, Boundary Lake, Koch Alberta, Pembina light sour, and Scotian light); four medium crude oils (Medium Gibson sour, Midale, Peace Pipe Sour, and West Texas Intermediate); eight heavy crude oils (Lloyd blend, Wabasca Heavy, Western Canadian Blend, Bow River North, Bow River South, Fosterton, Seal Heavy, and Sockeye 2000) and eight diluted bitumen (Access Western Blend, Cold Lake Crude, Western Canadian Select, Christina dilbit blend, Borealis Heavy Blend, Kearl Lake).

and alkylated) PAHs, which have been demonstrated to cause both acute and sublethal toxicity via numerous mechanisms of action that are important in delayed responses and in long-term residual and chronic effects, such as carcinogenesis, reproductive failures, developmental deformities, and immune suppression. A summary of PAH concentrations in some commonly transported crude oils and diluted bitumen is presented in the previous chapter (Table 2-2).

There is a wealth of toxicological literature describing the sublethal effects of commonly transported crude oil components.[29] The individual-level acute and sublethal effects include early-life-stage developmental defects, reduced growth and reproductive capacity, behavioral impairment, altered bioenergetics, genetic damage, impaired immune function and hence disease resistance, and enzymatic and hormonal changes including endocrine and hypothalamic-pituitary-adrenal axis implications, which can lead to population- and ecosystem-level impacts (e.g., changes to the base of food webs that affect higher-level consumers).

In addition to direct, chemically based mechanisms, crude oils can also result in acute and sublethal effects via physical mechanisms. The physical coating of biological surfaces impedes an organism's movement and can alter behavior and/or hamper respiration. An example of this would be the coating of gills and permeable skin surfaces of fish.

Toxicity of Diluted Bitumen

As noted in Chapter 2, many of the chemical compounds in diluted bitumen are found in other crude oils, and thus toxic properties are expected to be similar in many respects, although the relative proportions differ. The bitumen is a concentrate of the high-molecular-weight components of a conventional crude, while the diluent is a concentrate of the low-molecular-weight components. In contrast to other crude oils, very limited data exist on the toxicity of diluted bitumen, although much is known about the diluent components as they are commonly found in other crude oils. The potential for diluted bitumen to significantly weather, adhere to particles, submerge, and possibly sink in quiescent areas, coupled with its high content of recalcitrant resins and asphaltenes, can result in unrecoverable sunken oil and thus prolonged chronic exposure of benthic organisms.

Without a more detailed chemical characterization of this initial chemical pool, coupled with characterization of weathering or biodegradation products the comparison of diluted bitumen toxicity to other commonly transported crudes cannot be completed. As discussed in Chapter 2, a large fraction of diluted bitumen consists of an array of currently uncharacterized chemicals. This situation is not unique to diluted bitumen and applies to other crude oils; however, diluted bitumen has a larger number of unknown polar compounds. Furthermore, the adhesion of residual bitumen oil to biological surfaces may lead to physical coating of organisms (Figure 3-5), impacting movement, behavior, feeding, thermoregulation, and respiration.

There are very few published laboratory experiments specifically investigating the toxicity of diluted bitumen, despite its use and transport in North America for over 40 years. Currently, there is only one laboratory study investigating the toxicity of a diluted bitumen, working with larval fish (Japanese medaka, *Oryzias latipes*).[78] Sublethal concentrations of soluble components of the oil or "water-accumulated fraction" derived from Access Western Blend caused an increased prevalence of blue sac disease, impaired development, and abnormal swim bladders upon hatching. In addition, exposures resulted in various genetic markers for physiological stress that are commonly observed with exposure to other crude oils. It has been well established that exposure to conventional crude oils can cause embryo toxicity in fish primarily due to the 3- to 5-ring alkylated PAHs. Concentrations of 3- and 4-ring unsubstituted parent and alkyl PAHs are similar or higher in Cold Lake Blend in comparison to other commonly transported crudes as listed in Table 2-2 in Chapter 2, although that blend contains higher levels of total phenanthrenes, fluorenes and chrysenes compared with light and medium crude oils. Therefore, based on the limited available research, it is expected that exposure to diluted bitumen would cause similar or higher chronic toxicity to fish embryos, although further chemical characterization of diluted bitumen along with more toxicity studies in other fish species would be required to confirm this. Delayed effects from acute or chronic exposure or chronic toxicity studies of the residual bitumen component have also not been investigated. Other additional mechanisms of action and sub-lethal effects to other species and life stages have also not been studied for diluted bitumen.

While there have been few experimental dose-response investigations of diluted bitumen toxicity, the toxicity of diluted bitumen spilled into the environment has been inferred from post-spill field observations and investigations as well as laboratory studies using field samples from recent diluted bitumen spills. In addition, organisms collected near the oil sands deposits have been examined.[79] Natural bitumen deposits are exposed in the banks of rivers in the Athabasca oil sands region, and studies of wild fish from these locations have found sublethal biochemical and hormonal responses, including the classic response to exposures of PAHs, namely increased levels of ethoxyresorufin-O-deethylase activity and a reduction in steroid production in comparison to fish from reference areas.[79d] For early life stages of fish, these biochemical responses can be related to the observed deformities in embryos and larvae following exposure to waters affected by oil sands.[79c]

The significance of the field observations was confirmed by laboratory-based studies of sediment toxicity in water bodies of the oil sands region.[79a,79b] Compared to controls, fish eggs and embryos of fathead minnow (*Pimephales promelas*) and white sucker (*Catostomus commersonii*) showed increased mortality, reduced hatching success, delayed timing of hatching, abnormal development of embryos, and deformations and reduced size of larvae. The same effects have been observed during exposures to weathered conventional crude oils.

Bitumen contains several metals that are potentially toxic and are discussed in further detail in the human toxicity section below. However, potential bioavailability and toxicity of metals in the diluted bitumen transported by pipelines remains little studied. Studies in the vicinity of the oil sands mining sites in Alberta have documented increased concentrations of cadmium, copper, lead, mercury, nickel, silver, and zinc in snowmelt (reflecting atmospheric emissions and redeposition) and river waters, reaching levels of concern for the protection of aquatic life.[7b] Commonly measured metals, such as vanadium and nickel are found at higher levels in bitumen (and diluted bitumen) compared to other crude oils but these metals are predominantly bound in organic-metal porphyrin structures that are less bioavailable. Few data address this issue, however, and weathering and biodegradation processes have the potential to release these metals. Overall the toxicity (chronic and acute) of fresh and weathered diluted bitumen and its residues to freshwater, estuarine, and marine species at various life stages is at this time very understudied compared with other commonly transported light and medium crude oils. Furthermore, bioaccumulation and impacts to the food web and trophic transfer issues have not received attention for diluted bitumen in contrast to that of commonly transported crudes.

In the case of the spill of diluted bitumen into the Kalamazoo River (Box 3-1), toxicological effects on fishes were studied by a team from the U.S. Geological Survey within the framework of the Natural Resource Damage Assessment (NRDA).[80] Fish were sampled within a few weeks of the spill to obtain a gross pathological assessment of general health. When fish in oiled reaches were compared to fish in an upstream reference reach, significant adverse changes were evident. These differences were not observed in subsequent years.

The likelihood of submergence and sinking of weathered diluted bitumen, often as OPAs, merits particular attention because it presents distinct routes of exposure to the biota. In addition, the sunken oil may not be recoverable, thus resulting in protracted periods of exposure. Many aquatic animals also consume particles directly or indirectly from the bottom sediment, underwater plant or macroalgal surfaces, or the water column, which may include oil droplets and OPAs. Contact of droplets with respiratory surfaces (e.g., fish gills) or with permeable dermal surfaces can interfere with respiration. Organisms feeding on oil-contaminated material can in turn be consumed by predators, which can pass contaminants up the food web.

Human Health

The major ways in which humans may be exposed to chemicals following an oil spill are via (i) inhalation of volatile organics, (ii) dermal (skin) exposure through direct physical contact with the oil, (iii) ingestion of contaminated drinking water, and (iv) ingestion of contaminated food. This section aims to compare potentially hazardous characteristics of diluted bitumen with those of other commonly transported crude oils and identify whether diluted bitumen may pose similar or distinct human health risks.

Inhalation and dermal exposure are typically immediate concerns for the first responders and the public in the vicinity of the spill. Dermal exposure can be addressed by use of appropriate personal protective equipment, and so the focus here is on inhalation hazards and contamination of drinking water and food sources. While the issue of drinking water contamination can also be immediate, such concerns may persist for some time and may occur well beyond the initial site of the spill, depending on the fate and transport of the spilled oil. Similarly, contamination of agricultural produce (e.g., by irrigation water) and fisheries may be a concern in the near term. However, there may also be longer-term concern for food safety. During a spill, water bodies for human recreational use (swimming, boating, and fishing) are closed until deemed safe. For these reasons, the immediate human health issues are considered first and then the potential longer-term human health concerns are considered separately.

Initial human health concerns

During the initial days of spill response, the major components of concern to human health in crude oils include the volatile compounds—benzene, toluene, ethylbenzene, and xylenes (collectively called BTEX) and hydrogen sulfide (H_2S)—that can result in acute and sublethal effects via inhalation exposure. Benzene is also a well-known human carcinogen. Benzene is typically present in crude oils and is frequently monitored to assess both inhalation and drinking water supplies. Health and safety concerns regarding exposure to volatile organic compounds have also been addressed in the context of response (see Chapter 4). Oil spills can occur in populated areas and result in immediate human exposure; the 2013 Mayflower, Arkansas, oil spill in a residential area exemplifies the importance of an immediate, effective response to mitigate human health hazards (Box 3-5).

Table 3-1 shows the 5-year average BTEX concentrations and ranges for light, medium, heavy and diluted bitumen crude including the four representative crude oils described in Chapter 2.[74] The data indicate that the BTEX constituents in diluted bitumen (reported in % by volume) are not distinctly different from other crude oils. The average BTEX in diluted bitumen at 0.89 % vol was similar to the heavy crude oils at 0.84 % vol, whereas light and medium crude oils were 2.56 and 2.80 % vol respectively. The light crude oils category represented the highest variability. By comparison, the 5-year average values reported for all condensates listed in the crude monitor database averaged 4.22% BTEX by volume.[74]

Exposure to hydrogen sulfide (H_2S) gas is also of concern as it can immediately damage the central nervous system and act as a chemical asphyxiant.[81] Given that hydrogen sulfide is corrosive, its levels were discussed in the previous report.[2] Diluted bitumen typically contains similar or lower levels of hydrogen sulfide compared to the other crude oils.

Another immediate (and potentially longer-term) concern relative to human health is the contamination of drinking water in the vicinity of or downstream of the spill. The more soluble BTEX components, especially benzene, are of most concern for drinking water and are also discussed in Chapter 4. In water, the solubilities of BTEX compounds range from approximately 150 to 1,800 mg/L[82] making them significantly more soluble than most other hydrocarbons. The USEPA has established water quality standards for BTEX compounds that are regulated by the Safe Drinking Water Act (SDWA). USEPA's National Primary Drinking Water Regulations set maximum contaminant level (MCLs). For example, MCLs for benzene are 0.005 mg/L (5 parts per billion or ppb). In a 2015 pipeline spill of Bakken crude oil into the Yellowstone River in Montana, oil related volatile organic contaminants (e.g., benzene) were detected at levels of concern at a drinking water treatment plant forcing the closure of the water intakes. Given the similar concentrations of BTEX in diluted bitumen and other commonly transported crude oils in the U.S. pipeline system, it is not expected that the BTEX components in diluted bitumen will pose a higher immediate risk to human health during the initial spill response phase.

Of additional concern with respect to immediate human (and any exposed organism) toxicity is the presence of elevated concentrations of

---

Box 3-5
Mayflower, Arkansas: ExxonMobil

On March 29, 2013, the 20-in. Pegasus oil pipeline, constructed in 1947-1948 and operated by ExxonMobil Pipeline Corporation, ruptured in Mayflower, Arkansas, releasing 3,190 barrels of Wabasca Heavy crude oil in a residential area. The Pegasus pipeline is buried 24 in. underground at the release site. The oil flowed down the street and into a drainage ditch and tributary that led to Dawson Cove, an arm of Lake Conway. The upper part of Dawson Cove was heavily forested and flooded at the time of the spill. The lower part of Dawson Cove is separated from the lake by a road with open culverts. Response crews were on site within 30 min of detection of the release, including ExxonMobil Pipeline employees and federal, state, and local responders. Staff from the Arkansas Game and Fish Commission quickly constructed earthen berms at the head of Dawson Cove, which allowed the oil to be contained within the cove, with no oil documented as entering Lake Conway.

Wabasca Heavy crude oil is a blend typically composed of ~80% bitumen obtained by polymer injection and water flooding from the Athabasca region and 20% diluent; this blend typically has a 19-20 API gravity, ~1% BTEX, and 4.15% sulfur.[74]

Twenty-two homes were evacuated. Air quality monitoring in residential areas in the first week or so after the release reported benzene below detection (0.05 ppm) but volatile organic compounds (VOCs) of up to 29 parts per million (ppm) on the day of the spill and 3 ppm by the third day. Higher levels of both benzene and VOCs were measured in work areas, and workers wore respiratory protection as specified in the health and safety plan.

There were no reports of oil sinking in the quiet waters of Dawson Cove. However, the dense vegetation had to be removed to access the oil, and intensive mechanical methods were used to remove trees, shrubs, rootballs, and other or-

---

unknown volatile and/or water-soluble compounds. This concern is not unique to diluted bitumen and is of concern for all crude oils.[21] However, the concentrations of polar compounds such as those containing nitrogen are higher in diluted bitumen. Therefore, there may be chemicals of toxicological concern unique to diluted bitumen that have not yet been characterized. This consideration is an important caveat for an assessment of relative human health risk for diluted bitumen spills compared to spills of commonly transported crude oils: an assessment based only on known BTEX concentrations must be considered incomplete and therefore tentative.

ganic debris from about 14.5 acres of forested wetland. The intensive mechanical operations mixed oil into the soils in the cove area. The post-cleanup assessment found low levels of PAHs in the soils that were mostly below sediment toxicity concerns. Nonetheless, because of chronic sheening in the cove, in August 2014 ExxonMobil was required to (i) excavate 800 yd^3 of sediment from the 1,300-ft-long tributary upstream of the cove, and backfill as needed with clean sediment; (ii) place an organoclay soil amendment on 2 acres of sediment surface within the remaining vegetated area; and (iii) place a reactive cap on about 4.5 acres of open water.

This spill shows that quick response can be very effective. Diluted bitumen products will initially float with the lack of turbulence in fresh water, and at that time they can be cleaned to levels of low toxicity, but are persistent and can cause chronic sheening that can trigger the need for extensive treatment, particularly in inland areas where natural removal processes are slow.

Photo Credit: U.S. Environmental Protection Agency

Longer-term human health concerns

In addressing longer term human health concerns, protection of water supplies is a focus of spill response activities. For reasons described earlier in the report, weathered diluted bitumen has a greater potential to submerge or sink, presenting a greater potential for chronic contamination of a water supply that may result in a long closure time for drinking water sources. Another serious outcome in the case of incomplete removal of sunken weathered bitumen could be a longer lasting impairment of a surface-water source of drinking water.

In the United States, groundwater supplies about 32% of drinking water[83] and there have been studies of crude oil contamination of groundwater, such as the site in Bemidji, MN (Box 3-2). Given the potential for longer travel time of crude oil in groundwater systems, the impacts may be detected later than for surface waters and may be inherently more difficult to remediate than for surface water bodies. The environmental behavior of a diluted bitumen spill with respect to groundwater contamination is discussed above.

There are various delineations of groundwater zones in which limits are set for both residential and industrial zones[84] for acceptable levels of Total Petroleum Hydrocarbons (TPH) and BTEX components. In the recent diluted bitumen spill in Marshall, MI, over 150 drinking water wells were monitored for a variety of inorganic and organic oil (and non-oil) related chemical contaminants. The contaminants monitored included nickel, vanadium, and organic compounds, including BTEX. TPH in the gasoline range (GRO, ranging in carbon atoms from C_6-C_{10}) and the diesel range (DRO, $>C_{10}$-C_{28}) were measured. The public health assessment report released in 2013[85] stated that there were no detections of crude oil related chemicals in any of the sampled wells, other than iron and nickel. These elevated levels of iron and nickel were only found in a few of the over 150 wells sampled and were deemed to be a natural occurrence for the wells in Calhoun and Kalamazoo Counties. The absence of groundwater contamination along the river was not unexpected given that groundwater flows toward the river and its floodplain in almost all of the affected reach.[86]

When there is deemed to be a risk of human exposure via food consumption, the collection of fish and other food items (e.g., shellfish) may be prohibited over space and time based on sheening and/or the presence of targeted compounds in the tissues of food items. The main class of compounds that are measured for food safety is the polycyclic aromatic hydrocarbons (PAHs), although others, including metals may be monitored. As Table 2-2 in Chapter 2 shows, there are a number of higher molecular weight PAHs that the USEPA has listed as probable human carcinogens for a representative set of light, medium, and heavy crudes, in addition to diluted bitumen. Out of the 16 EPA priority PAHs listed, 11 have slightly higher to over 3 times higher levels of these carcinogenic compounds in Cold Lake diluted bitumen compared to the four other crude oils. For example, benzo[a]pyrene levels in Cold Lake diluted bitumen are 3.01 µg/g compared to the light and medium crudes ranging from 0.25–0.74 µg/g. As mentioned above there also are many unknown compounds in crude oils for which we do not know the environmental fate, bioavailability and potential impact to organisms. Further chemical composition details are required, together with assessments on bioavailability and toxicity to organisms. Furthermore, details on how and to what the chemical constituents are biodegraded when diluted bitumen is spilled into the environment or metabolized to by organisms that bioaccumulate oil constituents are also required for a full understanding of the toxicity of diluted bitumen.

Accumulation of oil-derived metals into the food chain and ultimately into food for human consumption is also possible. However, levels of metals are usually very low in crude oils and only a few are measured (typically, nickel, vanadium, copper, cadmium and lead). Levels of vanadium and nickel in the various crude oils are reported as these are usually the metals of highest concentration compared to other trace metals. Using the 5 year average levels of vanadium and nickel in the same oils as detailed for the BTEX comparison shows that diluted bitumen concentrations are over 7 and 2.5 times higher than in the example light and medium crude oils respectively. For example, five year average levels of nickel and vanadium in diluted bitumen are 60 and 152 mg/L respectively compared to 8 and 20 mg/L in light and 23 and 60 mg/L in medium crudes. The bioavailability and toxicity of metals can also be very dependent upon the specific form (speciation) of the metal, which is dependent upon a variety of environmental parameters, including redox status (related to the presence or absence of oxygen) and pH content, both of which can be modified due to an oil spill. However, these measurements do not take the bioavailability of these metals into consideration and they are likely to be less bioavailable given that they are commonly found in tightly bound organic structures (porphyrins) in diluted bitumen. Although evidence of food web contamination from a diluted bitumen spill is lacking, risks for benthic organisms and their consumers cannot be ruled out given the limited data available regarding chemical composition and how biodegradation and other weathering processes might change the chemical composition of the residual oil. A more complete understanding of the chemical constituents of all crude oils and their weathering products is necessary to support a more thorough toxicological assessment. Based on the limited available evidence for diluted bitumen, however, it appears that it would pose a lower or similar hazard to human health in the short-term for the chemicals currently monitored and assessed. However, the potential toxicological risks to humans and animals, particularly longer-term exposures, are currently unknown compared to commonly transported crude oils.

CONCLUSIONS

In some respects the environmental effects of diluted bitumen spills resemble those of spills of other commonly transported crude oils, as long as the diluent and bitumen remain mixed and in their original proportions. The movement of the oil on land and on the water surface, and its toxicity to wildlife and people, are similar at the outset. Once a spill occurs, however, exposure of diluted bitumen to the atmosphere allows the lighter diluent fraction to evaporate, resulting in residual bitumen that has several distinctive characteristics, being particularly dense, viscous, and with a strong tendency to adhere to surfaces and to submerge beneath the water surface and potentially sink to the sediments.

In light of these characteristics, diluted bitumen spills in the environment pose particular challenges when they reach water bodies. Progressive evaporative loss of the diluent leaves behind the relatively dense and viscous bitumen, which can then become submerged, perhaps first by adhering to particles, and ultimately sink to the sediments. The density of the residual oil need not exceed that of the water to submerge if conditions are conducive to the formation of oil-particle aggregates with densities greater than water, and this may be a common situation in inland and coastal waters where suspended particulate matter abounds. The loss of the lighter fraction and resultant potential for submergence of residual oil manifests more quickly and will involve a greater fraction of the spilled oil than in the case of light and medium crude oils. Toxicity of the residual bitumen has received little study, although toxic effects of both organic substances and associated metals have been observed in the vicinity of oil sands deposits in western Canada. The difficulty of recovering sunken oil and the recalcitrant nature of bitumen mean that aquatic biota may be exposed to the material for longer periods than in the case of lighter oils that sometimes sink to the bottom but are relatively biodegradable.

Impacts of diluted bitumen spills are expected to vary across the great diversity of inland and coastal water settings, which present varying scenarios for transport and submergence of the oil, for the feasibility of recovering oil that has sunk, for the nature of the aquatic biota that would be exposed, and for human uses of the water and water bodies. Flow and mixing patterns, turbulence, and the nature and abundance of natural particulate matter are among the most important considerations.

The toxicity of diluted bitumen has scarcely been studied using a direct, experimental dose-response approach in the laboratory, although it has been inferred from studies in surface waters draining the oil sands deposits, as well by post-spill sampling and bioassays. Evidence so far suggests that the BTEX in the diluent—as well as certain PAHs in the bitumen—can have toxic effects, but these same compounds also occur in other commonly transported crude oils. Many of the PAHs known to cause chronic and sublethal effects, such as alkyl PAHs and EPA priority PAHs, are at similar or higher levels than those of commonly transported crude oils. Diluted bitumen or its weathered residues may contain other, uncharacterized compounds with toxic properties, but this awaits further investigation. There may also be various metabolites produced from diluted bitumen components that are also currently unknown and uncharacterized for their toxicological consequences. Until there is more toxicological research specifically targeting diluted bitumen, the acute, chronic, sub-lethal and longer-term toxicities of diluted bitumen relative to conventional crude oils will be poorly known.

4

Spill Response Planning
and Implementation

INTRODUCTION

The identification of an oil spill triggers the mobilization of personnel and equipment to protect the health and safety of the public, and to detect, contain, and recover the spilled oil while minimizing impacts to communities and the environment. Many characterizations of risk include probabilistic considerations of the magnitude and frequency of the hazard itself; for example, in the case of hurricane storm surge, these factors are estimated based on historic data. The multiple factors that can contribute to spill occurrence can include unpredictable accidents resulting from human actions; for example, the 2007 diluted bitumen spill in Burnaby, British Columbia (Box 3-4), was the result of construction activities unrelated to pipeline operations.[37b] The vulnerability of communities and environments potentially affected can be assessed in advance, however, and these factors become a key component of spill response planning.

For crude oil spills from transmission pipelines, the corridor of potential release is at least fixed and the impact areas are predictable. The types of environments, communities, and facilities that could potentially be impacted, and their sensitivities and vulnerabilities, can thus be identified in advance and are essential elements of spill response planning. Where the characteristics of the right of way are such that it is more or less sensitive to spills of different materials (e.g., diluted bitumen versus medium or light crude oil), these factors can be considered in advance; however, the specific characteristics of the material may change over time in any transmission pipeline operations, particularly for pipelines transporting diluted bitumen.

The transport, fate, and effects of spilled oil depend not only on the characteristics of the oil but also on the environments and conditions at the time and place of the spill. The consequences of a spill of diluted bitumen into a stream at base flow will differ from those at flood conditions based on the effect of turbulence on suspended particle formation, the availability of sediment particles for adhesion, the extent of the riparian or floodplain zone at risk, and the length of stream affected within the first few days. Similarly, the public safety issues surrounding volatilization of lighter fractions would be different in a recreational community on a holiday weekend compared to midweek during low season. Every spill presents a unique combination of conditions. Responders on scene must use their experience to adjust the response plan to the circumstances that confront them.

Given this uncertainty regarding the magnitude and character of any specific incident, spill response planning for pipelines is based on the concept of the "Worst Case Discharge," which is the largest foreseeable discharge of oil, including a discharge from fire or explosion, in adverse weather conditions. This is calculated as follows:[113]

The Worst Case Discharge is the largest volume, in barrels (cubic meters), of the following:

1. The pipeline's maximum release time in hours, plus the maximum shutdown response time in hours (based on historic discharge data or, in the absence of such historic data, the operator's best estimate), multiplied by the maximum flow rate expressed in barrels per hour (based on the maximum daily capacity of the pipeline), plus the largest line drainage volume after shutdown of the line section(s) in the response zone expressed in barrels (or cubic meters); or

2. The largest foreseeable discharge for the line section(s) within a response zone, expressed in barrels (or cubic meters), based on the maximum historic discharge, if one exists, adjusted for any subsequent corrective or preventive action taken; or

3. If the response zone contains one or more breakout tanks, the capacity of the single largest tank or battery of tanks within a single secondary containment system, adjusted for the capacity or size of the secondary containment system, expressed in barrels (or cubic meters).

In addition, pipeline operators may claim prevention credits for breakout tank secondary containment and other specific spill prevention measures.

This chapter outlines the main elements of spill response planning relevant to the consideration of diluted bitumen, describing the types of plans developed under the National Contingency Plan (NCP) and considering potential protection priorities for diluted bitumen spills. The roles and responsibilities of various agencies and entities in the development of these plans, including the NCP process, are described in Chapter 6. The next section focuses on activities that occur during actual spills and how these may need to vary for spills of diluted bitumen compared to crude oils. The chapter concludes with a summary of the specific challenges for spill response planning and implementation presented by the transport of diluted bitumen in pipelines.

IMPLEMENTATION OF PLANS

Predicting the Behavior of Spilled Oil

In framing and scaling an actual incident, responders ask the following questions:

- What and how much was/is being spilled?
- Where will it go?
- What are the resources at risk?
- What are the likely impacts?
- What should be done to reduce these impacts?

The observed type and volume of the spilled product drives initial actions related to the safety of responders and the public, mobilization of equipment, and estimates of the crude oil's likely behavior and pathway in the environment. Any delay in access to accurate information about the composition and properties of the spilled product can significantly affect the effectiveness of the response. Access to such information is critical, but specific compositions of products being transported by pipelines are usually not promptly available from pipeline operators or from the sources of the products they transport, and often vary over time in a particular pipeline. Compositional information is particularly important for diluted bitumen because the types and concentrations of diluents vary in ways that strongly affect the behavior of the spill and thus response strategies. Safety Data Sheets (SDSs) for crude oils are usually generic and provide ranges in reported properties, such as density; they do not provide information that responders need, such as the specific type of crude oil, density after weathering over time, chemical composition, and adhesion (which is rarely provided), among others. Because of their generic nature, responders seldom obtain the data they need from an SDS. If the crude oil name is provided and the crude oil is a standard type, readily available databases, such as Crudemonitor.ca, the Environment Canada oil properties database, and the National Oceanic and Atmospheric Administration (NOAA) Automated Data Inquiry for Oil Spills (ADIOS) oil library, can be consulted to obtain some of these data. However, if the spilled material is a blend that does not have a standard composition, but rather may change significantly from batch to batch, these databases may provide incomplete or inaccurate information. In such cases, batch-specific information is needed.

Modeling for guiding response activities is typically done for short durations and, thus, differs from modeling conducted to evaluate long-term impact. NOAA ADIOS is designed to provide oil weathering information for only 5 days. In addition, the NOAA Environmental Response Division uses the General NOAA Operational Modeling Environment (GNOME) to obtain modeling results of oil transport on the water surface, and the main purpose of these results is assisting the Unified Command of a spill in making the appropriate response decision. Such models would need to run with a minimal number of parameters and to attempt to capture the salient features of the release in terms of direction and magnitude.

While existing oil spill models can be used for the response to diluted bitumen spills, the main parameters are typically calibrated to conventional oils. For example, the windage factor, which provides the transport speed of oil, is typically equal to 3% to 4% in the early stages of a conventional oil spill,[87] and it is later decreased further as the oil weathers and forms emulsions. For diluted bitumen, the residual oil density can increase rapidly with the evaporation of the volatile diluent components. Since diluted bitumen does not promote the formation of emulsions, the windage factor of a diluted bitumen is initially low (e.g., 3%) and is not expected to decrease further with time. Another challenge in using existing oil spill models for diluted bitumen is the lack of sufficient experimentally obtained data to calibrate the modules with diluted bitumen.

Health and Safety Concerns

Because volatile organic compounds (VOCs) can evaporate rapidly when crude oil is released to the environment, public and worker safety related to air quality must be considered during the early stages of the response.[88] Light crude oils, particularly those produced during hydraulic fracturing of shales (e.g., Bakken and Eagle Ford) can pose significant air quality and explosion risks early in the response. Because benzene is a known carcinogen that is present in many petroleum products, it has the highest ranking in terms of potential for exceeding occupational exposure levels and community-exposure guidelines.[89] However, there may also be concerns about and monitoring of other VOCs, hydrogen sulfide, and explosion hazards (Chapter 3).

When air quality is a particular concern, there may be the need to establish a Public Health Unit within the Planning Section of the Unified Command to develop criteria for evacuations and reoccupation by the public. In addition, the Site Safety Plan for responders will have to follow Occupational Safety and Health Administration (OSHA) and American Conference of Industrial Hygienists guidelines for VOCs in general, and benzene, toluene, ethyl benzene, and xylenes in particular, as well as for hydrogen sulfide and other contaminants of concern. There may also be a need for real-time measurements of concentrations in work areas and for use by workers of passive air-monitoring and dosimeter badges, which are sent for analysis in order to monitor exposure. A program of this kind was implemented during the recent spills of diluted bitumen in Marshall, MI (Box 3-1),[12] and Mayflower, AR (Box 3-5), where the oil spread to areas in close proximity to residential areas.

Effects on water quality can also be significant. Spills of crude oil that reach water bodies can result in either closure of the affected water body to public use or posting of advisories to avoid oiled areas until there is no longer a potential for exposure. Such closures and advisories are likely to be longer when the spilled oil sinks in the water body and generates chronic sheening. For example, the Kalamazoo River and a reservoir known as Morrow Lake were closed for nearly 2 years after the Enbridge Pipeline spill in July 2010 (Box 3-1). Drinking water intakes may be shut down until testing determines that the water is safe to use, or the raw water may require additional treatment such as aeration and carbon filtration, as was conducted during the 2015 spill of crude oil from the Bridger Pipeline in Glendive, Montana, into the ice-covered Yellowstone River.[90]

Cleanup Endpoints

Cleanup endpoints are the criteria against which the response actions are measured, to determine if the goals and objectives have been met. Cleanup endpoints generally are set for water, shorelines, and soils. For spills in coastal and marine habitats, cleanup endpoints are usually based on the following guidelines[91] rather than analysis of samples for measurement of the concentration of selected contaminants:

- No oil observed: not detectable by sight, smell, or feel;
- Visible oil but no more than background amounts of oil;

- No longer generates sheens that will affect sensitive areas, wild-life, or human health;
- No longer rubs off on contact; and
- Oil removal to allow recovery/recolonization without causing more harm than natural removal of oil residues.

Cleanup endpoints for a specific spill are developed through consensus among the stakeholders, which can include public health officials. Key considerations are the trade-offs between aggressive techniques that remove the oil but also cause additional damage versus less intrusive techniques that rely on natural processes to remove, dilute, or bury residual oil (collectively known as natural attenuation). Cleanup endpoints for inland oil spills tend to be more stringent than those applied to spills in the marine environment and often require the use of more intensive cleanup methods that carry a risk of increased ecological impacts[92] for the following reasons:[91]

- Inland habitats often lack some of the physical processes (such as waves and tidal currents) that can speed the rate of natural removal of oil residues after treatment operations are terminated and can affect smaller water bodies where there are slower rates of dilution and degradation.
- The direct human uses of inland habitats, such as for drinking water, recreation, industrial use, and irrigation, require a higher degree of treatment compared to marine environments to avoid human health and socioeconomic impacts.
- Spills in close proximity to where people live, work, or recreate often require treatment to a higher level.
- There may be large-scale differences in water levels during the response, causing oil to be stranded well above normal levels where it can pose hazards to wildlife as well as humans using these areas.
- Many states have sediment quality guidelines that must be met as part of the remediation phase after the emergency response is completed.

Table 4-1, which has been adapted from Whelan et al.,[92] lists guidelines for establishing cleanup endpoints for spills in inland habitats. Achieving consensus on cleanup endpoints for spills of diluted bitumen can be challenging if the crude oil sinks and continues to generate sheens in areas of high public use, or where the residual crude oil adhering to substrates is difficult to remove.

TABLE 4-1 Guidelines for Selecting Cleanup Methods and Endpoints for Different Inland Habitats

Basis for Treatment	Applicable Habitats	Treatment Methods	Example Primary[a] Cleanup Endpoints	Guidelines for No Further Treatment Determination
Protection of Public Health and Safety	• High public use areas • Residential areas • Groundwater supplies	• Whatever needed to remove threats: excavate, cut, flush, remove/replace	• No visible oil • No detectable oil (sight or smell)	• When oil residues are no longer a threat to human health and safety • Falls below threshold odor or exposure limits
Protection of Sensitive Resources and Habitats	Wetlands, bird nesting areas, T&E species habitat, wildlife refuges, national parks, other protected areas	• Gross oil removal using vacuum, skimming, manual removal using walking boards in soft substrates • Passive recovery of sheens	• No free-floating black oil or mousse on the water surface • No accessible oiled debris • No oil in sediments that are used for nesting, hibernating, grubbing for food	• Usually determined by resource manager or land manager experts • Case studies that show habitat damage from aggressive treatment • Particular sensitivity of a species or habitat • Inability to replace habitat
Removing Aesthetic Impacts in High-Use Areas	• Hard substrates such as bedrock, gravel, seawalls, riprap • Beaches • Vegetation • Debris	• Wipe, high pressure, high temperature flush, cut, remove/replace	• No visible oil • No more than 20% stain or coat	• Less aggressive removal during seasonal low-use periods could allow natural processes to work • Consider how long before the oil weathers • Public information campaign concerning remaining staining required

Continued

TABLE 4-1 Continued

Basis for Treatment	Applicable Habitats	Treatment Methods	Example Primary[a] Cleanup Endpoints	Guidelines for No Further Treatment Determination
Removing Contact Hazard (both humans and wildlife)	• Hard substrates such as seawalls, riprap, bedrock • Vegetation • Debris • Soil	• Wipe, flush, cut, sorbent barriers, remove/replace	• No longer rubs off on contact • No oil that rubs off on sorbents	• Consider how long before the oil will weather to a nonsticky stain or coat • Avoid excessive vegetation removal • Falls below known limits for hazards • Public information campaign concerning remaining staining required
Mitigating Persistent Sheens	• Rivers, streams, other flowing water bodies • Lakes, ponds, other standing water bodies • Seasonally flooded wetlands	• Acutely remove the major sources of sheens (excavate, dredge, flush, cut, remove/replace) • Passively contain/recover sheens with booms and sorbents	• No longer generates sheens that affect sensitive resources • No longer releases black oil or mousse during flushing operations • No longer generates black oil or mousse during high-water events	In low-use areas: • Consider seasonal use and processes (e.g., flooding that speed natural removal) In high-use areas: • Education on considerations between aggressive removal and chronic sheens • Site specific studies to assess receptor risk

Mitigating Intermittent Sheens (triggered by rainfall, temperature changes, etc.)	• Rivers, streams, other flowing water bodies • Lakes, ponds, other standing water bodies • Seasonally flooded wetlands	• Acutely remove the major sources of sheens (excavate, dredge, flush, cut, remove/replace) • Passively contain/recover sheens with booms and sorbents	• No longer generates sheens that affect sensitive resources	In low-use areas: • Education on considerations between aggressive removal and chronic sheens In high-use areas: • Education on considerations between aggressive removal and chronic sheens • Site specific studies to assess receptor risk
Mitigating Sediment/Soil Contamination	• Upland soils • River/lake bed sediments • Wetland sediments	• Acutely remove the gross contamination (excavate, dredge, flush, cut, remove/replace) • Passively contain/recover remobilized oil with booms and sorbents • In situ techniques such as aeration, tilling, phytoremediation, adding nutrients	• No visual oil greater than stain or coat • Does not release black oil when disturbed • Agriculture or pasture for human use may need a ppm endpoint	• High risk of erosion or excessive sedimentation • Unacceptable changes in surface topography • Avoid excessive change in sediment/soil quality, e.g., organic matter content, grain size • Potential permanent change to the habitat type e.g., wetland to open water

a"Secondary Cleanup Endpoints should include "Or, as low as reasonably practicable considering net environmental benefit."
SOURCE: Adapted from Whelan et al.[92]

TACTICS FOR DETECTION, CONTAINMENT, AND RECOVERY OF SPILLS OF DILUTED BITUMEN

Spills to Land

Spills on dry land, if detected early, are often readily contained and recovered before extensive contamination of soil or groundwater (Chapter 3). One recent study[93] found that diluted bitumen penetrated a sand column more slowly than light, medium, and heavy conventional crude oils, indicating diluted bitumen soaked into sandy substrates may be no more difficult to recover than other crude oils. Problems occur when a light crude oil is released underground and not detected for days to months, or when the release is into a highly permeable substrate. When a pipeline released light crude oil underground in a gravel-outwash plain near Bemidji, Minnesota (Box 3-2), the combination of both of these factors led to one of the most extensive (and studied) incidents of groundwater contamination.[68] A more recent example is the subsurface release of about 20,000 barrels of Bakken crude oil from a pipeline in agricultural land near Tioga, ND. The light crude oil penetrated more than 30 ft into the ground. Extensive excavation and treatment of soil is required and is expected to take 2 years to complete.

Recovery methods for spills on land include manual and mechanical removal followed by offsite disposal, burning of oil that is pooled on the surface or in depressions and ditches, high-temperature thermal desorption or incineration (followed by return of treated soils to the spill site once they meet endpoints), and bioremediation of residual oils after gross oil removal. Cleanup of land spills is usually completed in weeks to months. When groundwaters are contaminated, cleanup is far more challenging and can extend over decades of time, with associated high costs.

Spills to Water and Wetlands

Detection

Floating crude oil is detected mostly by aerial observations, ground and water surveys, and, depending on the spill size and characteristics, remote sensing. These methods are well established and effective for any floating crude oil. These methods fail, however, when the crude oil submerges or sinks. Methods of detection employed in such cases have included (i) sonar systems, (ii) underwater cameras and videos, (iii) diver observations, (iv) sorbents, (v) laser fluorosensors, (vi) visual observations from the water surface, (vii) bottom sampling, (viii) water-column sampling, and (ix) the combination of in situ mass spectrometry with autonomous underwater vehicles.[94] These methods are not well established, are relatively slow, often provide only a snapshot of a small area, and suffer from many limitations depending on conditions such as wave height, water depth and currents, water turbidity, and ability to detect buried crude oil. Disturbance of sediments with a disk on a pole, followed by observations of floating crude oil globules and sheen appearing at the surface, was the preferred method for field detection and mapping of submerged oil in the case of the diluted bitumen spill into the Kalamazoo River near Marshall, MI.

FLOATING OIL RESPONSE TACTICS

On Water Containment and Recovery

Because most crude oils that could be released from pipelines are expected to float initially, the first response actions are to deploy booms to contain the crude oil and protect sensitive areas, and use of skimming, vacuum, and sorbents to recover the contained crude oil. When booms are deployed quickly and well, a large amount of the floating crude oil can be recovered, particularly in streams and rivers where the crude oil is contained between the banks. However, there are many conditions where floating crude oil cannot be effectively contained and recovered, including high-flow and turbulent conditions in rivers; strong winds and large waves, especially in estuaries and coastal waters; coverage of a water body by snow and ice; and limited access, such as in remote areas, floodplains, other wetlands, and difficult terrain. Under these conditions, responders look for downstream or downcurrent locations where response equipment can be effectively deployed. On rivers, these can include impoundments, bridge crossings, and boat ramps. Temporary roads may have to be constructed to gain access, particularly to wetlands and small streams.

As diluted bitumen weathers and the diluent is lost by volatilization, the floating bitumen will become highly viscous and require specialized skimming and pumping systems capable of handling such high-viscosity oils. Mesocosm tests with Cold Lake Winter Blend (CLWB) with an initial thickness of 30 mm floating on water in outdoor tanks, where the oil increased in viscosity over time up to 30,000 cP, showed that conventional heavy oil skimmers were effective.[9c] However, under real-world conditions, depending on the source and temperature, weathered diluted bitumen can increase in viscosity up to 1,000,000 cP (see Chapter 2).[94] Moreover, the residue may not continue to float (Chapter 3), a possibility not addressed by most spill response plans that exist today.

Weathered diluted bitumen adheres strongly to shorelines, vegetation, and debris and will be more difficult to remove from these surfaces for on water recovery by flushing methods, compared to conventional crude oils. The adhered oil will also pose a threat of fouling of habitats and wildlife because it will more quickly weather into a viscous, sticky residue.

Dispersants

The efficacy of dispersants is related to the viscosity of the spilled crude oil. Based on laboratory[95] and mesocosm studies[9c] with the diluted bitumens CLWB and Access Western Blend (AWB), interactions with dispersants exhibit roughly the same dependence on viscosity. Accordingly, dispersants are moderately effective at 10,000 cP but have little to no effectiveness at viscosities >20,000 cP. Diluted bitumen, for which the viscosity thresholds are reached within 6-12 hours under mild to moderate open water conditions and at temperatures of 15°C to 20°C, therefore have a narrower window of opportunity for effective use of dispersants than conventional crude oils. In comparison, medium crude oils are expected to reach these thresholds within 24-72 hours in temperate conditions and possibly within 12-24 hours during the winter.[46a]

In situ Burning

Mesoscale tests[9c] showed that burning is viable on diluted bitumen weathered up to 1 day, with removal efficiencies of 50% to 75%. The burn residues were sticky and easily submerged; thus, there would need to be rapid removal of the residues to prevent sinking. In comparison, medium and heavy crude oils are expected to burn at 85% to 99% removal efficiencies over longer periods of weathering, thus generating much lower amounts of burn residue. Formations of stable oil emulsions containing >25% water are difficult to ignite and burn less efficiently,[96] regardless of the oil type. However, the mesocosm studies showed that the water uptake in both AWB and CLWB was as a mechanically mixed and unstable oil-water combination, and not as a stable, uniform emulsion (see discussion of emulsification in Chapter 3). Such combinations would likely break up during calm-water periods, which might increase the effectiveness of a burn.

Surface Washing Agents

Because diluted bitumen spills are expected to adhere strongly to surfaces, tests have been conducted using chemical agents designed to enhance crude oil removal and listed on the National Product Schedule under the heading of Surface Washing Agents. Studies have shown that

CLWB that had weathered on granite tiles under various conditions (on water, in sun, or in shade) could not be removed by low-pressure, ambient-temperature water, but could be removed by high-pressure, high-temperature flushing when used in combination with a surface washing agent after up to 5 days of weathering.[9c] Similar results were obtained during the response to the Burnaby spill (Box 3-4).[76] As in all spills, early application of surface washing agents increases their effectiveness. In fact, their use has been preapproved by Regional Response Team 6 since 2003.[97] During the Refugio spill, responders reported success in removing weathered oil from surfaces using dry-ice blasting, a technique that may also find application with surface cleaning of diluted bitumen (Box 3-3).

RESPONSE TO NONFLOATING CRUDE OIL AND ITS RESIDUES

When crude oil is suspended in the water column or sinks to the bottom, response tactics must change. There are no known, effective strategies for recovery of crude oil that is suspended in the water column, particularly where it occurs as droplets or oil-particle aggregates. Accordingly, the objectives are to track the suspended material and to predict where it may sink to the bottom. Nets with various mesh sizes and towed at varying speeds have been tested to determine the adhesion and leak rates for diluted bitumen and its residues.[98] Submerged material adhered to nets that extended to 0.5 m depth with minimal leakage at tow speeds of 0.3 m/s (0.6 knots) for fine and medium mesh sizes. When full, the nets weighed 25 kg/m^2, making them difficult to recover by hand, and 25% to 50% of the oil leaked out when the nets were removed from the water. The recovered material stuck so firmly that the nets could not be reused. Submerged material deeper in the water column was swept under the net at water flow rates of 0.3 m/s.[99] Similar results have been reported for heavy oils, indicating that the use of nets as a removal method for any type of oil suspended in the water will be of very limited effectiveness. Ideal conditions would be in low-flow, relatively small rivers or streams where the nets could be placed across the water body and readily replaced before they failed. In open water environments, submerged oil would first have to be located and the nets then deployed quickly and effectively, which seems unlikely.

Other tactics for removal of oil suspended in the water column include various types of filter fences, such as gabions (wire cages) stuffed with sorbents (usually "Oil Snare," a polypropylene adsorbent) placed on the bottom downstream from the release or snares attached to frames placed downstream. None of these tactics has been documented as effective.

Tactics for removal of sunken crude oil include suction dredge, diver directed pumping and vacuuming, mechanical removal, manual removal, sorbents, trawls and nets, and agitation/refloating. Suction dredging is a standard technique for removing sediments from the bottom of a water body, and it has been used to recover sunken crude oil during at least five spills and most extensively after the Enbridge pipeline spill in the Kalamazoo River (Box 3-1). This method generates large volumes of water and sediment that have to be treated and properly disposed of. Thus, it works best for surgical removal of small concentrated areas of sunken crude oil.

The method used most frequently for removal of bulk crude oil that has accumulated at the bottom of a water body is diver directed pumping and vacuuming. Divers can target the sunken oil and regulate the flow to minimize removal of ancillary water and sediment. The rate of pumping must be adjusted to prevent shearing and emulsification of the oil, and to effectively move highly viscous oils.

When the sunken crude oil is solid or semisolid, removal using an excavator, clamshell dredge, or other mechanical equipment can be effective and, under favorable conditions, generates little additional water or sediment for handling and disposal. This equipment is readily available and, if deposits are near shore, can be operated from land. Deployment from barges is also feasible, but there are depth restrictions (< 6 m), the equipment is large and heavy, and the rate of recovery is slow.

Where the sunken oil occurs in discrete patches, manual removal in shallow water by wading, or in deeper water by divers, can be effective and allow selective recovery of crude oil if visibility is adequate. However, it is labor intensive and slow, and requires specialized gear for diving in contaminated water and special procedures and supplies for decontamination of divers.

Where the sunken material consists of oil-particle aggregates, it may be possible to refloat the crude oil by agitation of the bottom. Agitation using rakes or similar tools, injection of water using water wands, and injection of air using equipment such as pond aerators were all used during the cleanup of the Enbridge pipeline spill in the Kalamazoo River (Box 3-1).[9a] The refloated crude oil was recovered using skimmers or sorbents. However, depending on the conditions, a significant amount of the crude oil or oiled sediment sinks back to the bottom. The agitation can also simply mix the crude oil more deeply into the sediment. Careful testing is needed to determine the effectiveness of these methods.

Sometimes, crude oil from sunken aggregates returns to the surface as the water warms and the oil becomes less viscous and is able to separate from the sediment; it is not likely that the increased temperature affects the oil density relative to the water density.[100] Gas bubbles released naturally from the sediment can also result in oil transport to the surface, through a process known as ebullition.[101]

WASTE MANAGEMENT AND DISPOSAL

Management and minimization of wastes are key challenges during response to a spill. Any activities that increase the volume of oily waste will have a large impact on cost. For spills where the crude oil initially floats, then sinks, the response team will be faced with the management and disposal of conventional waste materials, such as sorbents, protective gear, skimmed oil, oiled solids removed from land, oiled debris, and oily liquids, as well as any materials collected during detection and recovery of sunken crude oil. If the recovery of sunken crude oil involves methods such as pumping, vacuuming, or dredging, very large volumes of crude oil, water, and sediment will be generated, requiring separation into different waste streams for further treatment prior to disposal. For example, over 237,000 yd^3 of materials were removed from the Kalamazoo River and its floodplain in 2010-2014 after the Enbridge pipeline spill.[102]

Because the collected crude oil may either float or sink and the character of the waste stream will vary widely over time, decanting systems tend to be custom designed. Waste management is typically divided into three phases: (i) separation and treatment of solids, (ii) separation and treatment of liquids, and, where allowed, (iii) final polishing of liquids prior to release at the spill site. Where wastes can be treated on land, methods such as dewatering using geotubes (requiring a large footprint for the treatment area) and carbon treatment of water are used. Geotubes are sediment-filled sleeves of geotextile fabric. Where wastes can or must be treated at the site, a series of decanting tanks (onshore) or barges (on the water) is used, often with the goal of being able to discharge the treated water back into the spill site. Table 4-2 provides a summary of the effectiveness of selected response tactics for spills of conventional crude oils compared to spills of diluted bitumen.

CONCLUSIONS

Spills of diluted bitumen will initially float regardless of the water density; thus, the first response actions are similar to those employed after spills of conventional crude oil. However, as the diluted bitumen weathers, its properties change (see Chapters 2 and 3) in ways that can affect the response. The time windows during which dispersants or in situ burning can be used effectively are much shorter for spills of diluted bitumen than for spills of conventional crude oils. The strong adhesion of diluted bitumen to surfaces requires higher pressures and temperatures when using flushing techniques. Because it is already highly degraded, natural attenuation of residual diluted bitumen is less likely to be effective, which can trigger the need for more aggressive removal actions.

TABLE 4-2 Effectiveness of Selected Response Tactics for Conventional Crude Oils Compared to Diluted Bitumen in Seawater

Tactic	Light Crude	Medium Crude	Heavy Crude	Diluted Bitumen
Dispersant Effectiveness	50% to 90% up to 72 hr	10% to 75% up to 72 hr	0%[a]	~50% at 6 hr ~0% after 12 hr
In Situ Burning	99% at 96 hr	99% at 96 hr	90% at 96 hr	50% to 75% up to 24 hr; not effective after 96 hr
Removal of Oil Adhered to Substrates	Washing with low pressure, ambient temperature	Washing with higher pressure, and higher temperatures	Washing with high-pressure, hot water; may require use of surface washing agents; dry-ice blasting	Washing with high-pressure, hot water; may require use of surface washing agents; possibly dry-ice blasting
Waste Generation	Lowest because of high natural removal processes	Moderate	High	Potentially highest, if benthic sediment removal is required

[a]Based on review of laboratory dispersant effectiveness tests reported by Environment Canada in the online Oil Properties Database.
SOURCE: Environment Canada[31]

Most spill response tactics are based on the assumption that the crude oil will float. When a significant fraction of the spilled crude oil becomes suspended in the water column or sinks to the bottom, the response becomes more complex because there are few proven techniques in the responder "tool box" for detection, containment, and recovery. Recovery of sunken crude oil often generates large amounts of water and sediments that require complex logistics for handling, separation, treatment, and proper disposal of wastes. When sunken crude oil refloats spontaneously over a protected period, it can trigger the need for aggressive removal to mitigate the threats to water intakes, the public, fish, and wildlife. All of these threats are greater for spills of diluted bitumen than for spills of commonly transported crude oils. They drive the need for more complete removal of spills in inland areas where cleanup endpoints are usually more stringent.

Every spill presents a unique combination of materials and conditions. Better documentation of the behavior of diluted bitumen when spilled and of effective recovery methods is needed, so that the response community can benefit from these experiences.

5

Comparing Properties Affecting Transport, Fate, Effects, and Response

The preceding chapters have outlined differences between commonly transported crude oils and diluted bitumen in terms of compositions, properties, and likely fates in the environment. Using a potential spill from a transmission pipeline as a framework, the comparison can be extended to consider how those differences relate to the responses that would be necessary. Three potential products of a spill via transmission pipeline that are analyzed include (i) a spill of crude oil (commonly light or medium crude), (ii) a spill of diluted bitumen, and (iii) the residue produced by weathering of diluted bitumen, a few days after a spill. The third point is important because, relative to a light or medium crude oil, diluted bitumen is likely to produce a heavy residue both more promptly and in greater quantity.

POTENTIAL OUTCOMES AND LEVEL OF CONCERN

Three sets of hazards can be identified. The first relates to the transport or movement of spilled products in the environment. The second relates to the fates of those products (sinking, evaporation, persistence, etc.). The third relates to the effects of those products (impaired water quality, toxicity, air pollution, etc.). For each set of hazards, drawing on information summarized in Chapters 2 and 3, key properties were identified that could be related to the level of concern regarding that hazard. Finally, the levels of concern associated with each of the three products of interest were compared. Overall, this chapter represents an effort to distill the accumulated information, doing so in ways that are relevant to the regulatory framework.

TRANSPORT

Environmental hazards related to the release of diluted bitumen, and its eventual conversion to weathered diluted bitumen, are summarized and compared with other commonly transported crude oils in Figure 5-1. In Figure 5-1 the first column lists properties affecting transport, for example, adhesion and solubility. The second column graphically summarizes how commonly transported crude oils (CTC), diluted bitumen (D), and the residue resulting from weathering of diluted bitumen (WD) compare in terms of that property. The second column in the figure shows, for example, the density of commonly transported crude oils is lower than that of diluted bitumen and that the density of weathered diluted bitumen is considerably higher. The third column lists potential outcomes related to each property of interest. For example, density is an important factor in determining whether a product will submerge and move, for example, in a river or stream, in suspension, or as part of the bed load. Finally, the fourth and fifth columns tabulate the level of concern relative to that associated with a commonly transported crude oil. For example, the densities of commonly transported crudes and of diluted bitumen are so similar that the levels of concern regarding the related potential outcomes are approximately equal (noted in Figure 5-1 as "Same"). In contrast, the higher density of weathered diluted bitumen causes an increased level of concern ("More"). Columns 4 and 5 are color-coded, indicating whether a high or low value would exacerbate the risk associated with a potential outcome.

Crude oil that floats on water is transported by different mechanisms than crude oil that submerges, often sorbed onto sediments, and is transported in suspension or in the bed load of streams and rivers. The greater density of weathered bitumen results in a greater level of concern that weathered bitumen will become submerged in an aquatic environment (Chapter 3). Even in the first days of a spill, the greater adhesive properties of diluted bitumen compared to commonly transported crude oils result in a greater level of concern. This concern derives from impacts on wildlife and vegetation and from the associated public reaction, as volunteers mobilize to rescue contaminated wildlife, for example.

The greater level of concern for weathered bitumen also reflects the potential magnitude of the long-term effects of a spill that reaches a water body. Given the known composition of diluted bitumen, a much greater proportion of the material released can be expected to become denser than water and/or adhere to sediments, thereby sinking and entering

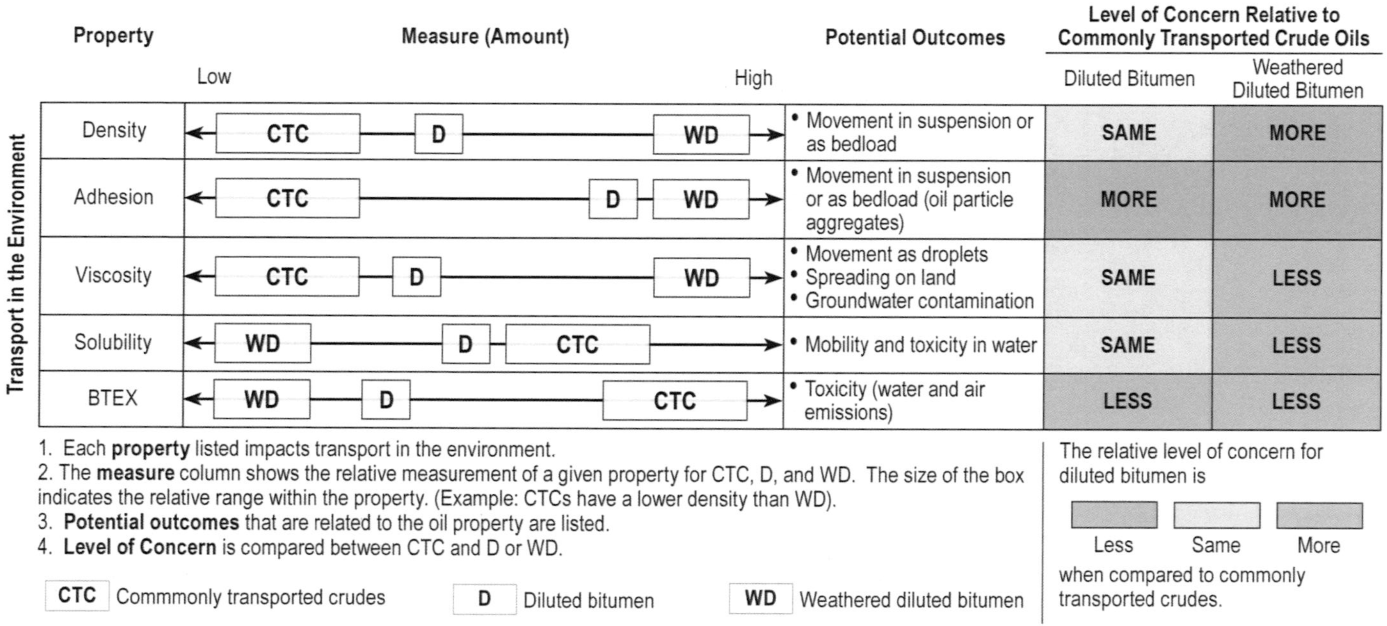

Transport in the Environment

Property	Measure (Amount)	Potential Outcomes	Level of Concern Relative to Commonly Transported Crude Oils	
	Low ——— High		Diluted Bitumen	Weathered Diluted Bitumen
Density	CTC — D ——→ WD	• Movement in suspension or as bedload	SAME	MORE
Adhesion	CTC ——— D — WD →	• Movement in suspension or as bedload (oil particle aggregates)	MORE	MORE
Viscosity	CTC — D ——→ WD	• Movement as droplets • Spreading on land • Groundwater contamination	SAME	LESS
Solubility	WD — D — CTC →	• Mobility and toxicity in water	SAME	LESS
BTEX	WD — D ——— CTC →	• Toxicity (water and air emissions)	LESS	LESS

1. Each **property** listed impacts transport in the environment.
2. The **measure** column shows the relative measurement of a given property for CTC, D, and WD. The size of the box indicates the relative range within the property. (Example: CTCs have a lower density than WD).
3. **Potential outcomes** that are related to the oil property are listed.
4. **Level of Concern** is compared between CTC and D or WD.

CTC Commmonly transported crudes   D Diluted bitumen   WD Weathered diluted bitumen

The relative level of concern for diluted bitumen is

Less   Same   More

when compared to commonly transported crudes.

FIGURE 5-1 Diluted bitumen relative to commonly transported crude oils: considerations related to transport in the environment. Acronyms: BTEX: benzene, toluene, ethylbenzene, xylenes.

the bed load and sediments of riverine, wetland, and coastal environments. Furthermore, once the weathered bitumen becomes incorporated into the bed load, it may be deposited some distance from the initial spill and remobilized in a future storm or flood. Thus, the benefits of being prepared to contain the diluted bitumen early during the response to a spill are substantial.

Very little is known about the risks associated with a subsurface release of diluted bitumen (i.e., into groundwater or a deep water column), particularly in terms of the risks of dissolution of the light, relatively water-soluble monoaromatics such as benzene, toluene, ethyl benzene, and xylenes (BTEX) into groundwater, where loss by volatilization and microbial degradation are likely to be slow.

The other properties noted in Figure 5-1, namely viscosity, solubility, and concentrations of the BTEX compounds, relate to other modes of transport and to toxicity. During the first few days after a spill, while substantial portions of the diluent are likely to remain, the diluted bitumen can be expected to be transported in a manner similar to that for commonly transported crude oil. The spill responses that are standard for crude oils can be applied effectively to diluted bitumen prior to weathering. On the other hand, once sufficient weathering has occurred, there is a lower level of concern that weathered bitumen will be transported by spreading compared to commonly transported crude oils or to diluted bitumen. Lower levels of concern also apply to other outcomes that are particularly important in spills of commonly transported crude oil, such as contamination of groundwater and release of toxic volatile compounds (e.g., BTEX).

FATE

Figure 5-2 summarizes estimates of the risks associated with a variety of fates for the spilled material. Those fates, or outcomes, such as sinking, surface coating, and burn residue are listed in the third column of the figure. The effects of density and adhesion on sinking and burial are similar to those on transport noted in Figure 5-1. The viscosity of diluted bitumen, or especially that of its weathered residues, however, is high enough that the risk of penetration in a soil profile is lower than that of a commonly transported crude oil.

While the diluent is retained, concentrations of the very lightest, most volatile, and most flammable hydrocarbons in a diluted bitumen (particularly if the diluent is a gas condensate) may be similar to or less than those of a light or medium crude. Accordingly, the risks associated with spills of diluted bitumen are approximately the same as those for spills of commonly transported crude oils. On the other hand, the weathered

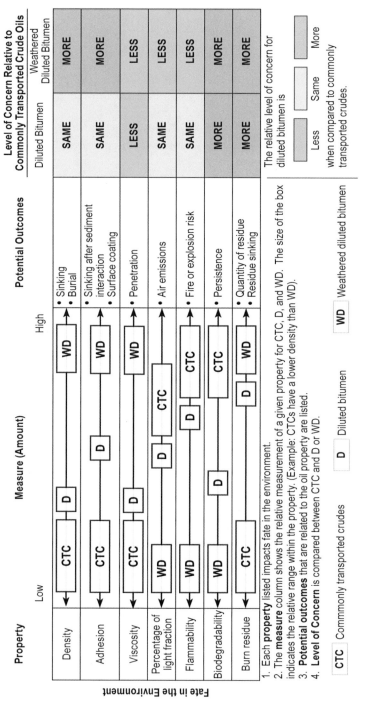

FIGURE 5-2 Diluted bitumen relative to commonly transported crude oils: considerations related to fate in the environment.

residues resulting from spills of diluted bitumen are less volatile and less flammable than those of commonly transported crude oils.

The presence in diluted bitumen, and particularly in its weathered residues, of large quantities of resins and asphaltenes heightens the level of concern about long-term persistence in the environment. Compared to commonly transported crude oils, a larger proportion of diluted bitumen and its weathered residues have the potential to become retained in sediments or in soils. Additionally, both the greater density and adhesive properties of weathered bitumen contribute to the greater likelihood of such an unfavorable outcome. If partially weathered bitumen becomes buried or entrained in sediments it can be more difficult to delineate and remove. This outcome can lead to a protracted period of exposures to the biota.

For commonly transported crude oils, there is a reasonable prospect that some portion of the oil not collected by cleanup activities will be biologically degraded in the water column or in soil and sediment. Figure 5-2 summarizes estimates of the risks associated with a variety of fates for the spilled material. Those fates, or outcomes, such as sinking, surface coating, and residue after burning, are listed in the third column of the figure. The effects of density and adhesion on sinking and burial are similar to those on transport noted in Figure 5-1.

Another fate issue is the comparison of the effectiveness of oil removal by in situ burning. Studies have shown[9c] the diluted bitumen has a shorter time window in which burning can be effective and a lower burn efficiency, compared to commonly transported crude oils. Thus, burning could leave a larger amount of residues for removal.

Overall, the level of concern for environmental persistence of a substantial portion of the released material is greater for diluted bitumen and much greater for weathered bitumen

EFFECTS

Effects, or outcomes, such as impaired water quality and hazardous air pollution, are considered in Figure 5-3. Density and adhesion again heighten the risks associated with diluted bitumen and especially with its weathered residues. Concentrations of the BTEX compounds in diluents are commonly high enough that related risks associated with diluted bitumen are similar to those with commonly transported crude oils, though the near absence of those compounds in the weathered residues yields a lowered risk.

For commonly transported crude oils, BTEX compounds are generally an immediate concern and this is true to a similar (or lesser) extent for diluted bitumen, which would correspond to the first phase of spill

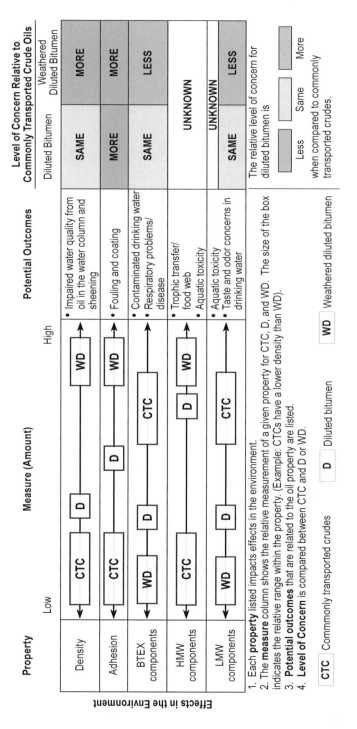

FIGURE 5-3 Diluted bitumen relative to commonly transported crude oils: considerations related to effects in the environment. Acronyms: BTEX: benzene, toluene, ethylbenzene, xylenes; HMW: high molecular weight; LMW: low molecular weight.

response. For the weathered bitumen BTEX is much less of a concern. In contrast, the level of concern is reversed for high molecular weight compounds that may be transferred in food webs and present a concern in terms of chronic toxicity. The major concern is the distinct lack of chemical characterization, biodegradation, and targeted toxicity studies for diluted bitumen compared to other commonly transported crude oils.

RESPONSE

As discussed in Chapter 4, a central goal of the response strategies for spills of crude oils is to protect human health and avoid or minimize long-term detrimental outcomes in the environment. To further address the statement of task, a comparative approach was used when considering the response to spills of diluted bitumen in relation to spills of commonly transported crude oil. In the comparisons presented in Figures 5-1 through 5-3, density and adhesiveness emerged as the most distinctive properties of diluted bitumen that could raise the level of concern for long-term environmental impairment, particularly due to the potential of some portions of a weathered bitumen to become submerged under the specific circumstances. Figure 5-4 illustrates that there are distinctive aspects of effective response techniques to spills of diluted bitumen that need to be addressed in the near term to avoid such an outcome.

The behavior of the light components of any spill in the first few days of an environmental release directly influences two aspects of the response. One immediate aspect of a spill is the potential health hazards to the public and the spill responders posed by volatile organic compounds. The extent of this risk for diluted bitumen products depends upon the diluent used; if condensate is used as a diluent, the level of concern may be comparable to that associated with light crude oil. The need to take appropriate measures, such as evacuation of nearby areas and/or providing appropriate personal protective equipment to the spill responders tasked with containing the spill, provides an added layer of complexity to response. These concerns are represented as being most significant in the first 2 days of the spill response for spills of both commonly transported crude oils and diluted bitumen and become less of a concern for diluted bitumen after weathering has occurred.

Another aspect of the diluent behavior that influences spill response during the first few days is the progressive loss of the diluent by evaporation, which decreases the tendency for diluted bitumen to float and spread on the surface of a receiving water body. This is noteworthy for diluted bitumen because the density of the oil being transported may be that of a medium crude oil, but this can change substantially as weathering occurs to yield a residual material with a density that approaches that of water.

97

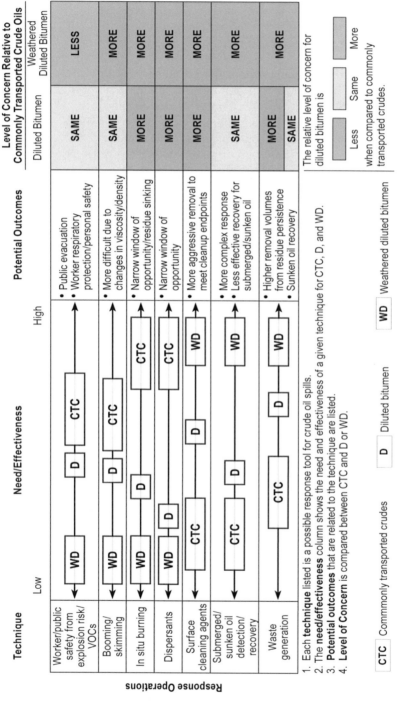

FIGURE 5-4 Response operations: diluted bitumen relative to commonly transported crude oils.

As a result, the initial period when diluted bitumen can be contained and recovered by established response protocols coincides with the period when the exposure risk due to volatiles influences the spill response activities. In particular, the assessment of where and how much diluted bitumen has been spilled may be held back if potentially dangerous levels of volatiles are encountered. As indicated in Figure 5-5, if containment, booming, and/or recovery of a large portion of the spilled diluted bitumen are not achieved during this initial period, a significant portion of the spilled oil may aggregate with particulate matter and become submerged. In contrast, in spills of commonly transported crude oils, the major fraction of the oil will likely continue to float for a longer period and often only a minor fraction may adhere to particulate matter and become submerged. Thus, the rapid loss of diluent inherently presents a distinct immediate challenge in responding to spills of diluted bitumen compared to spills of commonly transported crude oils.

Beyond the challenges presented by the behavior of the diluent, there are other distinctive aspects of the initial spill response for diluted bitumen. One is that use of in situ burning can only be effective for diluted bitumen within the first 24 hours (Figure 5-5). Unlike the outcome for commonly transported crude oils, where up to 99% can be removed by in situ burning over a longer period of time, a much lower percentage is removed upon burning diluted bitumen with the resulting burned residue being highly recalcitrant. Similarly, relative to commonly transported crude oils, there is a much narrower window of opportunity in which chemical dispersants can be applied effectively to spills of diluted bitumen. Based on laboratory and mesocosm studies, the effectiveness of dispersants on diluted bitumen is negligible after 6-12 hours, compared to effective dispersions of 50% to 90% after up to 72 hours for light and medium crude oils. A second distinctive aspect of diluted bitumen is that more oil may sorb onto structures and vegetation due to the increased adhesion described in Chapters 2 and 3. This will present challenges for removal with conventional methods.

If containment is not successful in the initial period of the response to a spill of diluted bitumen, the contrasts with spills of commonly transported crude oils are greatly amplified. As the diluted bitumen weathers, more of the oil may become submerged if the mixing energy is high and particulate matter is available. These products may either be carried downstream or be deposited in sediments when turbulence decreases. Under conditions where commonly transported crude oil may become submerged, in situ biodegradation can be considered when one is evaluating the impacts of the residual oil versus the impacts of intensive removal of the oil. However, biodegradation is less likely to be effective for a submerged and sunken weathered bitumen. The removal of this weathered

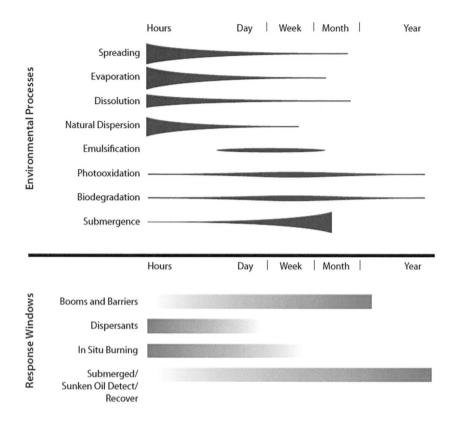

FIGURE 5-5 Time scales of environmental processes affecting spills of diluted bitumen and the windows of time for various response options.

bitumen may involve extensive physical disruption and generate large quantities of waste material and contaminated water to be treated. The environmental and economic benefits of avoiding this outcome are potentially great, but these benefits can only be achieved in the initial few days of the spill response. Thus, recognizing the need for prompt action while the diluent is still present is key to an effective response to a diluted bitumen spill. Based on this comparison of timelines and potential outcomes, there are distinct aspects of effective responses to spills of diluted bitumen in comparison to effective responses to spills of commonly transported crude oils.

CONCLUSIONS

The prospect of a release of crude oil into the environment through a pipeline failure inherently raises a number of concerns. These concerns include not only minimizing a number of possible long-term environmental impacts but also protecting the safety of responders and the public during and after the spill response. When all risks are considered systematically, there must be a greater level of concern associated with spills of diluted bitumen compared to spills of commonly transported crude oils.

In the context of fate, transport, and effects, the properties of diluted bitumen and weathered diluted bitumen that consistently result in greater levels of concern involve the higher density of the bitumen. The environmental outcome that should be most vigorously avoided in a spill response is the weathering of spilled diluted bitumen into heavy, sticky, sediment-laden residue that cannot readily be recovered, which requires dredging and disposal of large quantities of contaminated sediment and water, and which will not degrade if left in the environment. This weathering process begins rapidly following a release and can change the behavior of diluted bitumen in a matter of days. At the same time, the level of concern for responders and public safety associated with toxic and potentially explosive volatiles in the diluent fraction is similar as for commonly transported crude oils as these concerns are associated with properties of the diluents used.

The oil and pipeline industries and the response community have developed approaches for addressing releases of crude oil that are based on accumulated experience in responding to the diversity of spills that have occurred, as well as knowledge of the general properties of crude oil. This experience is predominantly based on spills of commonly transported crude oils that can be expected to float for some time. Given these greater levels of concern, spills of diluted bitumen should entail special immediate actions in response, for example, that the properties of diluted bitumen and weathered bitumen put such spills in a class by themselves.

6

Regulations Governing Spill Response Planning

This chapter addresses the fourth component of the committee's statement of task to "make a determination as to whether the differences between the environmental properties of diluted bitumen and those of other crude oils warrant modifications to the regulations governing spill response plans, spill preparedness, or cleanup." The first section of this chapter provides an overview of the federal framework for spill response planning, preparedness, and response. The second section examines how well this framework accounts for the unique characteristics of diluted bitumen and where improvements are warranted to improve its effectiveness. The chapter concludes that, in light of the committee's findings regarding the differences between diluted bitumen and commonly transported crude oils, modifications to the current regulatory framework are needed to better account for the unique characteristics of diluted bitumen.

FEDERAL SPILL PLANNING AND RESPONSE FRAMEWORK

In 1968, problems faced by officials responding to a large spill of oil from the tanker *Torrey Canyon* off the coast of England heightened awareness of the importance of effective spill planning.[103] That incident and later spills spurred recognition that the U.S. needed a coordinated approach to cope with potential spills in U.S. waters. Ultimately, a comprehensive system of spill reporting, containment, and cleanup was developed and codified in the National Oil and Hazardous Substances Pollution Contingency Plan, more commonly called the National Contingency

Plan (NCP). The NCP,[103] issued under the authority of the Comprehensive Environmental Response, Compensation, and Liability Act (CERCLA),[104] provides a multiagency federal blueprint for responding to oil spills and to releases of other hazardous substances.

Congress has broadened the scope of the NCP since its inception. A major milestone was the Oil Pollution Act of 1990 (OPA 90).[105] Enacted in the wake of the *Exxon Valdez* spill, OPA 90 expanded reporting, planning, and response requirements for oil spills, prompting extensive revisions of the NCP that were finalized in 1994.

Under the NCP, the planning and response process has several elements that are implemented at the pipeline or facility level and through national, regional, and area planning mechanisms. Figure 6-1 provides an overview of these elements and how they interact. Under ideal circumstances, planning at the pipeline or facility level should not occur in isolation, but as part of an integrated system that involves multiple governmental entities and the public.

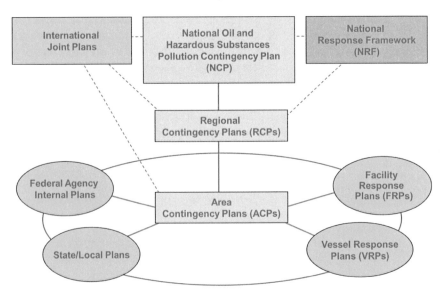

FIGURE 6-1 The relationship among the various levels of oil and hazardous materials response plans under and related to the U.S. National Contingency Plan. SOURCE: Adapted from U.S. Environmental Protection Agency[106]

National Response Team

The National Response Team (NRT)[107] is responsible for broad coordination and oversight of preparedness, planning, and response across the federal government. The U.S. Environmental Protection Agency (USEPA) and the U.S. Coast Guard (USCG) serve as chair and vice chair respectively, of the NRT. All federal agencies with some involvement in planning and response are members.

Regional Response Teams

There are 13 Regional Response Teams (RRTs)[108] in the United States, each representing a particular geographic region (including the Caribbean and the Pacific Basin). RRTs are composed of representatives from field offices of the federal agencies that make up the NRT, tribal nations, and representatives of state government. The major responsibilities of RRTs include response, planning, training, and coordination. Regional Contingency Plans are established by RRTs to ensure that the roles of federal and state agencies during an actual incident are clear, and also to identify resources that are available from each federal agency and state within their regions, including equipment, guidance, training, and technical expertise, for dealing with chemical releases or oil spills.

Area Contingency Planning

An Area Contingency Plan (ACP)[109] is a reference document prepared for the use of all agencies engaged in responding to environmental emergencies in a defined geographic area. ACPs are generally initiated by RRTs and can be developed based on geographic features and jurisdictional boundaries. Within the boundary of an ACP, subareas with unique circumstances that warrant tailored response strategies can also be defined. ACPs are designed to ensure that all responders have access to essential area-specific information and to promote interagency coordination as a means of improving the effectiveness of responses. Among other things, ACPs are a potential vehicle for identifying in advance spill scenarios that may damage areas that are environmentally sensitive or of special economic or cultural importance. ACPs can also pinpoint high-risk locations such as fixed facilities or pipelines. This can lead to developing response strategies that are effective in mitigating or preventing a substantial threat of discharge and ensuring that adequate resources (personnel, equipment, and supplies) are available for response to spills and releases with potential for serious environmental or other consequences.

Facility Response Plans

A Facility Response Plan (FRP) demonstrates a facility's preparedness to respond to a worst case oil discharge. Under the Clean Water Act,[110] as amended by OPA 90, preparation and submission of FRPs are generally required for land-based facilities that store and use oil, and for pipelines and vessels transporting oil if spills from these entities can reasonably be expected to cause "substantial harm"[111] to the environment, for example, by discharging oil into or on U.S. navigable waters.

Under the original terms of Presidential Executive Order 12777 issued in 1991,[112] and subsequent updates, responsibility for spill response planning is divided among four agencies—the Pipeline and Hazardous Materials Safety Administration (PHMSA), the USEPA, the USCG, and the Bureau of Safety and Environmental Enforcement (BSEE):

a. PHMSA has responsibility for overseeing preparation and approval of response plans for spills from onshore pipelines.
b. USEPA reviews and approves response plans for spills from non-transportation-related onshore facilities.
c. USCG performs these functions for vessels and onshore marine facilities.
d. BSEE in the U.S. Department of Interior oversees spill response planning for offshore facilities.

Some facilities fall under the jurisdiction of two or more federal agencies and must meet multiple requirements. For example, a complex may have a transportation-related transfer area regulated by USCG, a pipeline regulated by PHMSA, and a non-transportation-related oil storage area regulated by USEPA.

Significantly, the allocation of responsibility for responding to oil spills is different from the allocation of responsibility to oversee response planning. Authority for cleanup and response actions for all onshore oil spills is shared by USEPA and USCG. USCG leads responses to spills in the coastal zones and the Great Lakes, whereas USEPA has the lead for inland oil spills. However, while overseeing *planning* for spills from pipelines, PHMSA plays no role in the *response to such spills*, for which USEPA and USCG are responsible.

Onshore Pipeline Spill Response Plans

The response planning requirements applicable to operators of onshore pipelines are set forth in PHMSA's Part 194 regulations,[113] adopted in 1993 following passage of OPA 90. Under these regulations, each response plan must include procedures and a list of resources for responding to a Worst Case Discharge and to a substantial threat of such a discharge. To comply with this requirement, an operator is permitted to incorporate by reference into the response plan the appropriate procedures from its manual for operations, maintenance, and emergencies. Plans must be consistent with the NCP and the applicable ACP and must (i) determine the Worst Case Discharge for each of the operator's response zones; (ii) provide procedures and a list of resources for responding to a Worst Case Discharge and a substantial threat of such a discharge; (iii) ensure, by contract or otherwise, the availability of equipment and other response resources sufficient to address a Worst Case Discharge within the specified time limits; (iv) identify environmentally and economically sensitive areas; (v) designate the qualified individuals responsible for implementing the plan within each response zone; (vi) describe responsibilities of the operator and of relevant agencies in the event of a discharge and in mitigating or preventing a substantial threat of a discharge; and (vii) establish procedures for testing of equipment, training, drills, and other measures to ensure that the plan will be effectively implemented in the event of a spill.

Under Part 194, response plans must be updated immediately to address new or different operating conditions or information and resubmitted to PHMSA within 30 days. In addition, most plans must be revised and resubmitted every 5 years.

As part of their integrity management programs under PHMSA's Part 195 regulations,[114] pipeline operators must determine if a release from a pipeline segment could impact high consequence areas (HCAs). Identification of HCAs for hazardous liquid pipelines focuses on populated areas, drinking water sources, and unusually sensitive ecological resources, defined as follows:

- *Populated areas* include both high-population areas (called "urbanized areas" by the U.S. Census Bureau) and other populated areas (referred to by the Census Bureau as a "designated place").
- *Drinking water sources* include surface water or wells where a secondary source of water supply is not available. The land area in which spilled hazardous liquid could affect a water supply is also treated as an HCA.
- *Unusually sensitive ecological areas* include locations where critically imperiled species can be found, areas where multiple examples of federally listed threatened and endangered species are found, and areas where migratory waterbirds concentrate.

The purpose of identifying HCAs under Part 195 is to focus attention on pipeline maintenance, corrosion prevention, and other safeguards of pipeline integrity in areas where spills could have high impacts. Identification of HCAs is not intended to strengthen spill response plans under Part 194 although it clearly has value in designating environmentally and economically sensitive areas during preparation of response plans.

WEAKNESSES OF THE CURRENT PLANNING AND RESPONSE FRAMEWORK IN ADDRESSING SPILLS OF DILUTED BITUMEN

There are a number of areas where the current regulatory framework is not effectively addressing the potential environmental impacts of spills of diluted bitumen. This section reviews these shortcomings and highlights opportunities for the responsible agencies to improve their policies and procedures.

Because spills have unpredictable consequences and each spill is unique, spill response plans are an important starting point for effective response actions but do not provide an exact step-by-step protocol for responding to any specific incident. Nonetheless, while midcourse corrections and adjustments are unavoidable after a spill occurs, a good response plan forces facilities to rigorously assess in advance where spills might occur and what response strategies and resources must be in place to maximize an effective response. A good response plan also provides a means for communication and joint decision making among all the relevant entities and affected members of the public in advance of and during response to a spill. This communication in turn enables informed and effective collaboration and timely public engagement when a spill occurs. For this reason, there is a focus not just on response plans in isolation but has examined the relationship between plans and other aspects of the planning and response process. Similarly, it has looked not only at the role of PHMSA but at the interactions of the key agencies within the federal response structure.

Adequacy Review versus Checklist Approach

Roughly 400 pipeline response plans have been approved by PHMSA. The agency presently has approximately five members of its staff engaged in plan review, a more than doubling of its previous level. PHMSA's review of response plans is conducted in its Washington, DC, headquarters, with little or no involvement by its field staff. USEPA and USCG, by contrast, generally review plans for facilities within their jurisdictions in their regional offices. This enables staff members with relevant, hands-on cleanup experience and familiarity with the environments of the region to advise their colleagues who are reviewing the plans. In some USEPA offices, On-Scene Coordinators (OSCs) actually conduct the reviews themselves. When this occurs, the involvement of field staff familiar with the pipelines or industrial facilities and with HCAs within a region makes it more likely that the plan will identify and anticipate the greatest threats resulting from a spill.

At PHMSA, the review of plans is focused on completeness, using the Part 194 regulations as a checklist to ensure that all necessary components are present. Assuming the plan is complete, PHMSA's long-standing position is that it is legally obligated to approve the plan, and that it has no discretion to evaluate its likely adequacy and effectiveness or to recommend improvements. By contrast, USEPA and USCG review plans in two stages, the first focusing on completeness and the second on adequacy. As a result, reviewers from these agencies would be more likely to identify elements of a plan that may not be adequate when implemented for particular types of oils or in specific areas, and to request improvements.

Response plans submitted to PHMSA vary considerably in length and level of detail, based on the philosophy and goals of the particular pipeline operator. PHMSA takes no position on the length and detail of plans, providing they satisfy the minimum requirements of Part 194.

These aspects of the process at PHMSA make it unlikely that plans for pipelines that may transport diluted bitumen will be reviewed to assess whether they fully reflect planning needs relative to the properties of the diluted bitumen, the resources and expertise necessary to deal with spills of diluted bitumen, and the possible impacts of these spills on environmentally sensitive areas. A review process with a greater focus on adequacy and more interaction with regional response experts would create stronger incentives for plans to anticipate and include strategies for addressing the challenges posed by spills of diluted bitumen.

As noted above, unlike USEPA and USCG, PHMSA plays no role in coordinating or implementing actions taken in response to spills from pipelines. Thus, while PHMSA oversees development of plans and has some ability to enforce requirements pertaining to those plans, the measures actually necessary to address spills when they occur are outside its jurisdiction. This means that PHMSA is not in a position to bring to bear the lessons and experience from spill mitigation and response in its oversight and strengthening of planning. For example, PHMSA's minimal presence in the response to the spill in Marshall, MI (Box 3-1), limited its opportunity to learn key lessons from USEPA OSCs and other responders who encountered unanticipated cleanup challenges presented by diluted bitumen and its weathered residues.

The Part 194 regulations allow, but do not require, PHMSA to consult with USEPA and USCG during the review of response plans.[113] The normal practice is not to engage these other agencies during these reviews, although the Enbridge response plans were informally shared with USEPA following the Marshall, MI spill. Routine involvement of USCG and USEPA personnel in reviews by PHMSA would increase the likelihood that plans realistically anticipate pipeline and area-specific challenges, including those presented uniquely by diluted bitumen.

Strengthening the Area Contingency Planning Process

Area Contingency Plans augment facility response plans by enabling responsible officials from the relevant federal agencies to prioritize the greatest threats resulting from spills and releases within the area. Additionally, they facilitate development of strategies and identification of resource needs for mitigating environmental impacts resulting from these events. While USEPA and USCG are actively involved in the ACP process, participation by PHMSA is more limited, reflecting its lack of a direct role in managing response activities as well as its constrained resources. When PHMSA does participate, it is typically represented by regional staff and not by the headquarters team responsible for review and approval of plans.

The ACP process provides a vehicle for focusing attention on the challenges associated with particular types of crude oil such as diluted bitumen, the ecosystems and resources at risk from spills of these products, and the adjustments in response strategies necessary to mitigate environmental impacts. However, ACPs could do more to take advantage of this opportunity. ACPs often include Geographic Response Plans that identify and prioritize sensitive areas for protection, and some of them include site-specific protection strategies. However, nearly all of these plans were developed to respond to floating oil, and many have not been updated for years.[115] Because of the increased risks to water-column and benthic resources, there is a need to revise these plans to address spills of oils that have the potential to submerge or sink.

Under Part 194, plans must identify environmentally sensitive areas that could be affected by a pipeline spill.[113] PHMSA does not examine the completeness of this portion of a plan. The identification of such areas could point to water bodies and related ecosystems where the potential for diluted bitumen or its residues to sink and attach to sediments would be greatest and could also describe the related ecological consequences and cleanup challenges. A second concern is whether pipeline operators fully integrate HCAs (declared under Part 195 integrity management programs) with the description of sensitive areas in Part 194 plans, and whether either set of regulations is focusing attention on the specific risks posed to HCAs by different types of crude oils. Fuller collaboration between PHMSA and other agencies through the ACP process would provide the input to PHMSA and pipeline operators necessary to strengthen and focus these elements of response plans.

Identifying the Type of Crude Oil Being Transported and Its Properties

The Part 194 regulations require response plans to specify the type of oil transported by the pipeline.[113] In practice, however, operators generally meet this requirement by a generic description such as "crude oil" and PHMSA does not request greater specificity. Thus, the response plan will not indicate whether the pipeline is handling diluted bitumen and, if so, the source of the bitumen, the nature of the diluent, and physical properties such as density and viscosity. Such information is of great and immediate value to the response team addressing a spill. In the initial days of the diluted bitumen spill in Marshall, MI, confusion about the nature of the crude spilled and its relevant properties delayed an effective response because the team did not foresee the consequences of weathering and the resulting potential for residues of the diluted bitumen to submerge.

In addition to the response plan itself, the Safety Data Sheet (SDS) submitted by the pipeline operator is potentially a vehicle for identifying the type of crude oil and its properties. In conjunction with the plan and other information sources, a detailed SDS containing the pertinent information would assist responders setting near-term priorities directly following a spill of diluted bitumen. It would also assist the public in understanding the nature and consequences of the spill. The Part 194 regulations recommend but do not require that response plans include SDSs for the crude oil being transported by the pipeline section.[113] As a result, the adequacy of the SDS is not a required focus of the plan approval process by PHMSA and, in some cases, the agency may never receive the operator's SDS. SDSs found in pipeline spill response plans are typically generic and do not identify and differentiate specific crude oils and their properties. At present, however, there is no mechanism at PHMSA to require more informative SDSs.

Accurate and specific information identifying the product being transported and its properties must be readily available to responders and affected members of the public at the time of a spill. A more detailed response plan and SDS, along with additional steps to provide analytical data for the crude as soon as possible after the spill occurs, can help meet this need.

Relative to the PHMSA Part 194 process, the USCG plan-review framework is more focused on differentiating between types of crude oil.[114] USCG requires vessel operators to classify the crudes they are transporting based on gradations defined by specific gravity. Under this scheme, crude oils are assigned to one of five groups. Group II-IV oils are generally positively buoyant; Group V covers oils that are generally negatively buoyant (specific gravity equal to or greater than 1.0). The status of diluted bitumen under this scheme is ambiguous because as transported its specific gravity is below 1.0 but, after weathering, its specific gravity could approach or increase above this level. USCG is considering whether adjustments should be made to its classification system to more precisely address diluted bitumen. The present system is imperfect but provides a better mechanism for identifying the type of crude being transported than the generic descriptor used in plans submitted to PHMSA. If improved, the Coast Guard system might be adopted by PHMSA and other agencies as a uniform approach for classifying crude oils for spill response and planning purposes.

Certification of Oil Spill Removal Organizations

The USCG classification system plays an important role in its system for certifying Oil Spill Removal Organizations (OSROs), who contract with pipeline or facility operators to provide response resources and expertise on their behalf in the event of a spill. USCG certifies OSROs based on a policy document called the OSRO Classification Guidelines.[116] Certification is voluntary under the OSRO Classification Guidelines but OSROs often seek certification because reliance on a certified OSRO streamlines the level of detail required in response plans. The intent of the OSRO Classification Guidelines is to assess the OSRO's capabilities based on the type of oil it expects to address during a spill. Although USCG recognizes that floating and nonfloating oils (based on API gravity) require different equipment, the currently used (2013) OSRO Classification Guidelines do not specify the different response resources necessary to respond to non-floating oils that may become nonfloating during weathering, nor do they address diluted bitumen specifically.[i]

The Part 194 regulations allow pipeline operators to designate OSROs certified under the USCG OSRO Classification Guidelines to carry out response activities on their behalf. In practice, however, PHMSA does not review the OSRO's qualifications and capabilities beyond confirming that it has been certified by USCG for "Worst Case Discharge-1, Rivers and Streams." As a result, there is no independent inquiry into whether the OSRO has the expertise and resources to handle particular types of crudes such as diluted bitumen. If PHMSA instituted a qualitative plan

---

[i] As of the finalization of this report, the USCG is currently working on a classification for potentially nonfloating oils (Group IV) to include in the next release of its OSRO Guidelines.

review process instead of its present checklist approach, it would be able to confirm that the OSRO is in fact qualified for cleanups of the specific crude being transported by the pipeline. This would be helpful to ensure that, through the OSRO or otherwise, the pipeline operator has put in place response resources that are adequate to address spills of nonfloating oils such as diluted bitumen.

The combination of improved OSRO Classification Guidelines, greater specificity in identifying crudes oils in PHMSA-required plans, and more review of the adequacy of these plans could lead to a closer examination of the qualifications of OSROs identified in those plans to address unique characteristics of crude oils like diluted bitumen.

Updating Response Plans

The Part 194 regulations provide for updating of response plans at 5-year intervals and within 30 days where necessary to address new or different operating conditions or information. Following PHMSA's renewed enforcement authority in their 2011 reauthorization bill and in the aftermath of the Marshall, MI spill, PHMSA reviewed all response plans between 2012 and 2014. Based on the currently used review proto-col, only a handful of the more than 400 plans required substantial revi-sions. In hindsight, this would have been a useful opportunity to require pipelines transporting diluted bitumen to strengthen their plans. Present response plans dealing with aquatic environments focus on floating oil and do not address residues that submerge or sink. Updated plans could incorporate new technologies and methods, or faster response times, which are important for efficient response to spills of diluted bitumen.

Strengthening Drills and Exercising of Plans

An important element of spill preparedness is announced and unan-nounced "exercising" of response plans to determine how well they work in practice and to identify vulnerabilities and opportunities for improve-ment. It does not appear that these exercises (which involve to varying degrees USEPA, USCG, and PHMSA) have systematically simulated the unique challenges presented by spills of diluted bitumen. Announced and unannounced plan exercises need to devote greater attention to scenarios involving spills of diluted bitumen in order to increase readiness and capability to respond to diluted bitumen spills.

CONCLUSIONS

The current system for response planning, preparedness, and mitigation is geared to the properties, behavior in the environment, and response challenges of commonly transported light and medium crude oils. Thus, the focus is on preparing for spills of oil that float on the surface. Spills of diluted bitumen raise different issues because of the greater density that the product acquires as the diluent evaporates. This leads to the potential, depending on environmental conditions, for the diluted bitumen residues to aggregate with particles in the water column and submerge or sink to the bottom of the water body. Different strategies, expertise, and response capabilities are needed to effectively address this spill scenario. However, the current relevant U.S. regulations and agency practices do not capture the unique properties of diluted bitumen or encourage effective planning for spills of that product.

The PHMSA Part 194 regulations are critical to the preparation of thoughtful and effective response plans but, in their current form, do not focus pipeline operators on the properties of diluted bitumen and the associated response challenges or provide the information necessary for responders to address spills of diluted bitumen. Shortcomings of the regulations include the absence of any requirement to identify the type of crude oil being transported and its properties, to describe how ecologically sensitive areas might be impacted by spills of diluted bitumen, and to demonstrate that response strategies, resources, and capabilities are in place for effective responses to spills of diluted bitumen.

These weaknesses are compounded by PHMSA's reliance simply on a checklist approach to review plans, in contrast to focusing on the adequacy and effectiveness of plans, as is the case with reviews by USCG and USEPA. PHMSA also does not regularly consult with these agencies during plan reviews, despite their hands-on experience with spill response and knowledge of local conditions that could contribute to informed judgments about the adequacy of a plan. PHMSA has further been slow to require updates of plans, thus missing an opportunity for incorporating new information about the properties and environmental behavior of diluted bitumen in plans and enhancing their effectiveness.

The response plan is only one element in the overall federal system of preparedness and response. Other aspects of this system also need to address for the challenges imposed by diluted bitumen and to improve readiness. For example, a greater focus on diluted bitumen in area contingency planning, with PHMSA playing a more active role in the process, may help address this challenge. In addition, a uniform nomenclature system used by all agencies would help in the differentiation of diluted bitumen from other crudes.

7

Recommendations

As this report demonstrates, diluted bitumen has unique properties that differentiate it from commonly transported crude oils. Because of these properties, diluted bitumen's behavior in the environment following a spill is different from that of the light and medium crudes typically addressed in spill response planning, preparedness, and response. Of most significance are the physical and chemical changes that diluted bitumen undergoes as a result of weathering. When the diluent component volatizes, the remaining bitumen becomes denser and, depending on circumstances, may aggregate with particles in the water column and remain in suspension or sink to the bottom of a water body. The submergence of persistent residues of diluted bitumen in aquatic environments, as was seen in the Marshall, MI spill, and the potential for long-term deposition in sediments and banks and remobilization in the water column present environmental concerns and cleanup challenges not presented by commonly transported crude oils. These challenges necessitate different response strategies, including immediate efforts to recover spilled diluted bitumen before significant weathering occurs and effective methods to identify, contain, and recover suspended and sunken oil.

The existing framework for pipeline spill planning, preparedness, and response is generally designed to address floating oil and not residues that mix throughout in the water column, aggregate with particles, and sink to the bottom of aquatic environments. As a result, the pipeline operators and the agencies responsible for spill planning and response may not be adequately prepared for diluted bitumen spills and may lack the tools for effective cleanup. This is in part a shortcoming of the Pipeline and Hazardous Materials Safety Administration (PHMSA) Part 194 regulations and in part a shortcoming of the broader interagency contingency planning and response system. A more comprehensive and focused approach to diluted bitumen across the federal oil spill response family is necessary to improve preparedness for spills of diluted bitumen and to enable more effective cleanup and mitigation measures when these spills occur. The recommendations presented in this chapter are designed to achieve this goal.

Oil Spill Response Planning

Recommendation 1: To strengthen the preparedness for pipeline releases of oil from pipelines, the Part 194 regulations implemented by PHMSA should be modified so that spill response plans are effective in anticipating and ensuring an adequate response to spills of diluted bitumen. These modifications should

a. Require the plan to identify all of the transported crude oils using industry-standard names, such as Cold Lake Blend, and to include Safety Data Sheets for each of the named crude oils. Both the plan and the associated Safety Data Sheets should include spill-relevant properties and considerations.

b. Require that plans adequately describe the areas most sensitive to the effects of a diluted bitumen spill, including the water bodies potentially at risk.

c. Require that plans describe in sufficient detail response activities and resources to mitigate the impacts of spills of diluted bitumen, including capabilities for detection, containment, and recovery of submerged and sunken oil.

d. Require that PHMSA consult with the U.S Environmental Protection Agency (USEPA) and/or the U.S. Coast Guard (USCG) to obtain their input on whether response plans are adequate for spills of diluted bitumen.

e. Require that PHMSA conduct reviews of both the completeness and the adequacy of spill response plans for pipelines carrying diluted bitumen.

f. Require operators to provide to PHMSA, and to make publicly available on their websites, annual reports that indicate the volumes of diluted bitumen, light, medium, heavy, and any other crude oils carried by individual pipelines and the pipeline sections transporting them.

g. Require that plans specify procedures by which the pipeline operator will (i) identify the source and industry-standard name of any spilled diluted bitumen to a designated Federal On-Scene Coordinator, or equivalent state official, within 6 hours after a spill has been detected and if requested, (ii) provide a 1-L sample drawn from the batch of oil spilled within 24 hours of the spill, together with specific compositional information on the diluent.

These recommended changes in the Part 194 regulations would take important steps to ensure that spill response plans recognize the differences between diluted bitumen and commonly transported oils and that pipeline operators and agency responders have the special expertise, capabilities, and cleanup tools necessary for effective responses to spills of diluted bitumen.

Critical to effective planning and response is determining whether a pipeline segment is expected to transport diluted bitumen and, if so, identifying its cleanup-relevant properties (e.g., density, adhesion, viscosity, and biodegradability) before and after weathering. This should be accomplished by requiring spill response plans to provide industry-standard names[i] for the crude being transported, accompanied by a relevant properties description. This information would also be required in the Safety Data Sheet (SDS), which would be submitted to PHMSA for review as part of the response plan. The SDS is an important vehicle for communicating response information to affected persons who may not receive the formal response plan, such as local communities, fire departments, and medical personnel. A clear and specific description of the crude being spilled is critical to perform this informational function.

The response plans should also demonstrate that the operator understands the unique properties and potential environmental impacts of diluted bitumen and is prepared to implement response strategies that address its challenges. This should take the form of enhanced plan sections describing in detail (i) the areas most sensitive to the effects of a diluted bitumen spill, including the water bodies potentially at risk, and (ii) response strategies and resources necessary to mitigate the impacts of spills of diluted bitumen, including capabilities for detection, containment, and recovery of submerged and sunken oil.

The regulations should provide that PHMSA will review these plan elements not simply to determine whether the plan is complete (the current "checklist" approach) but also to examine whether the plan is adequate and effective in anticipating and preparing for a diluted bitu-

---

[i] The industry-standard names should be agreed by both the industry and the relevant government regulators and responders.

men spill. This approach would be a significant change in how PHMSA interprets its responsibilities under OPA 90. Thus, PHMSA would need to reexamine its legal authority to determine whether it has discretion to conduct an adequacy review of plans submitted by pipeline operators.

The PHMSA regulations should also provide that, in conducting such an adequacy review, PHMSA will routinely share the plan with USEPA and USCG and obtain their feedback. This consultation will take advantage of the on-the-ground spill response expertise and experience of these agencies (which PHMSA lacks because it has no direct role in the response process), thus bringing it to bear in examining the adequacy of plans in preparing for diluted bitumen spills.

Requiring submission of annual reports by pipeline operators documenting the volumes of various crudes being transported and the pipeline routes and sections carrying them will fill a fundamental gap in publicly available information and enable responders and communities to be more knowledgeable about the types of spills that might impact their areas and the potential consequences. This will facilitate better planning at the area, regional, and national levels and encourage more informed public engagement.

Finally, while the identification of diluted bitumen and other crudes by industry-standard names should be sufficient in the response plan and SDS, more detailed compositional information will likely be needed when a spill occurs, both to guide the emergency response as well as over the longer term to support forensic chemistry evaluations and site remediation. Given that compositions of oils carried in pipelines typically vary over time, and in the case of diluted bitumen the diluent may be particularly variable, there should be an expedited procedure for characterizing the specific composition of spilled crude oil after a spill. Crude oils transported by pipelines are routinely sampled and the samples are temporarily archived until the shipments are delivered to their final destinations in case questions about the quality of the delivered oil arise at refineries. The regulations should require that, in the event of a spill, sample(s) of crude oil from the archived set representing the spilled oil will be made available within 24 hours to the responsible government agency if requested. The 24-hour time period includes transit time from the origin of the line, which may be thousands of miles away. A minimum volume of 1-L should be sufficient for chemical analysis. Pipeline operators already maintain a custodial sample and the processes are already in place for the collection and storage. The chemical analysis from the crude oil sample will give the composition of the crude oil being transported, which will benefit critical early decisions responding to spills of crude oil. In addition, the operator should be required to inform the relevant authorities of the source and industry-standard name of any spilled diluted bitumen within 6 hours of a spill being detected. The 6-hour time period was selected as a reasonable balance between the usefulness of the information to the response efforts and the ability of an operator to obtain the information. Because of the potential for submergence within a relatively short window, it is critical that this information be provided to the response community with enough time to act. On the other hand, because transmission pipelines operate in batch shipments, it will take some time for the operator to conclusively identify the specific batch that was spilled.

Oil Spill Response

Recommendation 2: USEPA, USCG, and the oil and pipeline industry should support the development of effective techniques for detection, containment, and recovery of submerged and sunken oils in aquatic environments.

Spills of diluted bitumen products where the crude oil submerges in the water column or sinks to the bottom are particularly difficult for responders. Most of the effective response methods are based on the premise that the spilled oil floats. Proven methods are needed so that responders have effective means by which to determine where the crude oil is, track its movement over time, and effectively recover it. Detection of diluted bitumen spills on the bottom may pose different challenges than for conventional nonfloating crude oils because the diluted bitumen can occur as oil-particle aggregates that require different detection methods than those used to detect bulk crude oil on the bottom. In situations where water is moving, there are no proven techniques for containment of suspended or sunken crude oil to prevent remobilization and spreading prior to recovery. Various techniques have been proposed but few have been shown to be effective. Once the crude oil has sunk to the bottom, recovery methods are selected based on the environmental setting, amount and distribution of the crude oil, and cleanup endpoints. Better technologies are needed to minimize water and sediment removal and improve the separation and treatment of oil, water, and sediment.

Recommendation 3: USEPA, USCG, and state and local agencies should adopt the use of industry-standard names for crude oils, including diluted bitumen, in their oversight of oil spill response planning.

A common nomenclature for identifying diluted bitumen and other crudes should be used across the federal response family and by state and local responders to improve communication and ensure that responders have accurate and detailed information about the composition of the crude oil being spilled. This is particularly important for diluted bitumen, which has unique and potentially variable properties not fully recognized by the federal response community that need to be clearly identified at the time of a spill. In addition, a common nomenclature system will have benefits for spills of other crude oils as well. This system of product names, which would be incorporated in response plans under Recommendation 1, should be developed through a collaborative process among PHMSA, USEPA, USCG, and state and local agencies and then be adopted by all agencies.

USCG Classification System

Recommendation 4: USCG should revise its oil-grouping classifications to more accurately reflect the properties of diluted bitumen and to recognize it as a potentially nonfloating oil after evaporation of the diluent. PHMSA and USEPA should incorporate these revisions into their planning and regulations.

USCG categorizes oils into groups defined by their specific gravity, with Group IV oils defined as having a specific gravity equal to or greater than 0.95 and less than 1.0 and Group V oils as having a specific gravity equal to or greater than 1.0. These groups are important because they are used in Oil Spill Removal Organization classifications and setting of guidelines for response capabilities. Under the current approach, diluted bitumen oil products would be classified as Group IV because the fresh oil has a specific gravity less than 1.0. However, once spilled, weathering and loss of the diluent will result in a bitumen with a specific gravity that could approach or become greater than 1.0. The 2013 OSRO Guidelines lack guidance for nonfloating oils in terms of response and transportation and the USCG is in the process of creating a nonfloating oil classification system. Some oils, such as diluted bitumen, have unique characteristics that may cause an evolution from a Group IV oil to a potentially submerged oil (even if not surpassing a specific gravity of 1.0, as in the case of oil-particle aggregates formed in fresh water). A revised USCG classification system can better address these types of oils and should be incorporated in USCG, USEPA, and PHMSA regulations and strategies.

Advanced Predictive Modeling

Recommendation 5: The National Oceanic and Atmospheric Administration (NOAA) should lead an effort to acquire all

data that are relevant to advanced predictive modeling for spills of diluted bitumen being transported by pipeline.

A combination of information from oil property and behavior testing, as well as from oil spill models, can improve the response to an oil spill. Accurate spill modeling is now a very important part of both contingency planning and actual spill response. Spill models combine the latest information on oil fate and behavior with hydrodynamic modeling to predict where the oil will go and how much it will change before it arrives. The movement of crude oils is predicted by using the water current and wind speeds along the predicted water path. In addition to predicting the movement, these models can estimate the amount of evaporation, the possibility of emulsification, the amount of dissolution and subsequent movement of the dissolved component, the amount and fate of the portion that is naturally dispersed, and the amount of oil deposited and remaining on shorelines.

In the United States, much of the behavior and weathering information for spills in water is provided by NOAA via their Automated Data Inquiry for Oil Spills (ADIOS) oil weathering model. The NOAA Office of Response and Restoration, which developed and maintains ADIOS, has a mandate to support oil spill response, whereas other agencies do not have the base funding to do the same kind of work. This oil spill response tool models how different types of oil weather in the marine environment. The success of NOAA's ADIOS is due to their extensive database of more than a thousand different crude oils and refined products. However, the NOAA databases would benefit from additional information on all diluted bitumen products. Data on diluted bitumen that are relevant to the NOAA databases are typically found in bulk form and are currently available through Environment Canada databases and other resources. NOAA should lead an effort to fill this gap.

Improved Coordination

Recommendation 6: USEPA, USCG, PHMSA, and state and local agencies should increase coordination and share lessons learned to improve the area contingency planning process and to strengthen preparedness for spills of diluted bitumen. These agencies should jointly conduct announced and unannounced exercises for spills of diluted bitumen.

Improved coordination and communication among the many agencies with spill responsibilities would be valuable in creating stronger awareness of the response challenges posed by diluted bitumen, its unique properties, and the most effective strategies and resources for addressing diluted bitumen spills. There are several vehicles for enhanced collaboration but the area contingency planning process is particularly important because it encourages a focus on the pipeline routes, types of crude being transported, and sensitive water bodies and ecosystems potentially impacted by spills in a defined area. Through the ACP process, plan reviewers and responders can share information gleaned from diluted bitumen response actions, better integrating these two aspects of the response effort. Strong PHMSA participation in area and regional planning, which generally has not occurred up to now, is essential for this collaboration to succeed.

The committee's understanding is that plan exercises, which are critical to evaluate plan adequacy and responder preparedness, have devoted little if any attention to diluted bitumen spills. USEPA, USCG, and PHMSA together with state and local partners should work together to ensure that announced and unannounced exercises include diluted bitumen spill scenarios so that agencies and pipeline operators can obtain feedback and experience regarding the adequacy of plans for these spills and improve response capabilities.

Improved Understanding of Adhesion

Recommendation 7: USEPA should develop a standard for quantifying and reporting adhesion because it is a key property of fresh and weathered diluted bitumen. The procedure should be compatible with the quantity of the custodial sample collected by pipeline operators.

As highlighted by Table 2-6, diluted bitumen, and particularly its weathered residues, are highly adhesive. The amount of diluted bitumen residue that will adhere to a clean needle is more than 100-fold greater than the amount that adheres when a clean needle is immersed in the residue of a weathered light, crude oil. Reduction of uncertainties about the chemical cause of diluted bitumen's avid adhesion, and development of a method that quantifies adhesion precisely, would be useful in tailoring optimal cleanup procedures.

A limited study of the mechanism of adhesion, perhaps by analysis of the materials that adhere to surfaces (as opposed to analyses of the whole oils), should be used to inform the development of a method of adhesion measurement that will provide information not otherwise available to the spill response community and which can be well standardized.

PRIORITY RESEARCH AREAS

As outlined throughout this report, many differences between diluted bitumen and commonly transported crudes are well established. While there are clearly enough data and information to support the findings and recommendations outlined herein, a more comprehensive understanding of diluted bitumen and its properties, environmental, and human health effects, would improve spill response in the future. There still remain areas of uncertainty that hamper effective spill planning and response. These uncertainties span a range of issues, including diluted bitumen's behavior in the environment under different conditions, its detection when submerged, and the best response strategies for mitigating the impacts of submerged oil. Because of their importance to spill planners and responders, a concerted effort to fill these knowledge gaps with additional research is essential.

Further research is needed to better understand the behavior of diluted bitumen in the environment, including consideration of the diversity of environmental settings in which spills could occur, the chemical constituents and their toxicological effects, as well as to develop more effective methods for detection and recovery of spilled diluted bitumen, particularly after it becomes submerged or sunken in water bodies. Some of these research needs have been articulated in the past, including a report from the National Coastal Research Council (on behalf of USCG) entitled "Spills of Nonfloating Oils: Risk and Response," which included specific recommendations for detection, monitoring, modeling, and recovery of submerged oil; however this report focuses mainly on marine environments and did not consider the particular characteristics of diluted bitumen. The recent and projected future escalation of diluted bitumen transport in pipelines (and by other modes) has increased the possibility of release, and therefore research on the effects of diluted bitumen spills on the environment has become an ever more pressing need. Major questions targeted for research include the following:

Transport and fate in the environment. How will the various combinations of bitumen and diluent change (weather) upon release, and can we predict when submergence is likely under a variety of conditions (turbulence, suspended matter, contact with benthic surfaces, plants, etc.)? How does biodegradation influence toxicity, for example, by "releasing" previously bound chemical constituents or by producing more toxic biodegradation products?

Ecological and human health risks of weathered diluted bitumen. While much is known about the toxicity of hydrocarbon components in the commonly used diluents (e.g., benzene), there has been little study of the potential toxicity of bitumen to people or wildlife. For submerged and sunken oil in particular, there may be routes of exposure that have not been considered sufficiently, such as sensitive egg and larval life stages of fishes, and exposures may occur over protracted periods given the apparent resistance of bitumen components to biodegradation. Other indirect effects of the oil may relate to enhanced bioavailability of co-occurring pollutants and altering properties of the impacted ecosystem (i.e. redox status, dissolved oxygen levels and pH).

Detection and quantification of submerged and sunken oil. The Marshall, MI oil spill demonstrated that current options for sunken oil quantification are either unreliable (e.g., total petroleum hydrocarbons) or very expensive and time consuming (chemical fingerprinting), and consequently "poling" to disturb the sediments and observe the resultant appearance of floating globules and sheen became the primary means of mapping sunken oil (Box 3-1). Better measurement techniques should be a research priority.

Techniques to intercept and recover submerged oil on the move. Submerged oil moving downstream in rivers or following wind- or tidally driven currents could be intercepted in theory, but in reality no techniques are known to be efficacious to capture oil beneath the water surface. Research should strive to develop options for the diversity of environmental settings in which oil can be spilled.

Alternatives to dredging to recover sunken oil. Dredging is costly and environmentally destructive, producing voluminous waste that often must be landfilled, and therefore alternatives should be sought. Agitation and collection of resultant floating oil was conducted in the Kalamazoo River, but its efficacy in a particular spill needs investigation before being deployed again. Other alternatives should be studied as well.

These research priorities are targeted broadly to the research community, but a specific mention is needed regarding the role of local and regional scientists in spill response.[117] Improved access and collaboration with these scientists would help advance the scientific understanding of how oil behaves in the environment, particularly for emerging issues such as spills of diluted bitumen. Scientists from outside the formal response framework are typically not included in the formal oil spill response activities and, as a result, are often barred from site access by response officials, and their requests for source materials are denied. This situation hinders fundamental research on spill events—research that should ultimately benefit spill planning—and may also provide immediate benefit to response officials.

FINAL THOUGHTS

Diluted bitumen has received extensive publicity in the past 5 years and will continue to be of interest due to production from Canadian oil sands. As more diluted bitumen is transported, the need for efficient spill response planning, preparedness, and cleanup will be increasingly important. It is difficult to be completely prepared for a potential spill of diluted bitumen because our experience is limited to just a few significant spills, the products involved can vary in chemical composition, and the environmental settings where spills could occur are extremely diverse. Nonetheless, the recommendations put forward are designed to improve current oil spill planning and response to reduce negative impacts on human health and the environment.

References

1. Pipeline Safety, Regulatory Certainty, and Job Creation Act. Public Law 112-90, 2011.
2. National Research Council, *Effects of Diluted Bitumen on Crude Oil Transmission Pipelines*. The National Academies Press: Washington, DC, 2013.
3. Transportation and Housing and Urban Development, and Related Agencies Appropriations Bill. Senate Report 113-45, 2014.
4. Levine, S.; Taylor, G.; Arthur, D.; Tolleth, M. *Understanding Crude Oil and Product Markets*; American Petroleum Institute: Washington, DC, 2014.
5. (a) Canadian Association of Petroleum Producers *Crude Oil Forecast, Markets & Transportation*; Publication Number 2015-0007; 2015; (b) Crosby, S.; Fay, R.; Groark, C.; Kani, A.; Smith, J. R.; Sullivan, T.; Pavia, R., Transporting Alberta Oil Sands Products: Defining the Issues and Assessing the Risks. *U.S. Dept. of Commerce, NOAA Technical Memorandum NOS OR&R 43. Emergency Response Division, NOAA* **2013**, 153.
6. POLARIS Applied Sciences Inc. A Comparison of the Properties of Diluted Bitumen Crudes with other Oils. http://crrc.unh.edu/sites/crrc.unh.edu/files/comparison_bitumen_other_oils_polaris_2014.pdf.
7. (a) Dupuis, A.; Ucan-Marin, F., A Literature Review on the Aquatic Toxicology of Petroleum Oil: An Overview of Oil Properties and Effects to Aquatic Biota. *DFO Can. Sci. Advis. Sec. Res. Doc.* **2015**, *007*; (b) Gosselin, P.; Hrudey, S. E.; Naeth, M. A.; Plourde, A.; Therrien, R.; Van Der Kraak, G.; Xu, Z., *Environmental and Health Impacts of Canada's Oil Sands Industry*. The Royal Society of Canada: Ottawa, Canada, 2010.
8. Environment Canada; Fisheries and Oceans Canada; Natural Resources Canada, *Properties, Composition and Marine Spill Behavior, Fate and Transport of Two Diluted Bitumen Products from the Canadian Oil Sands*. Environment Canada: Ottawa, Canada, 2013.

9. (a) Fitzpatrick, F. A.; Boufadel, M. C.; Johnson, R.; Lee, K.; Graan, T. P.; Bejarano, A. C.; Zhu, Z.; Waterman, D.; Capone, D. M.; Hayter, E.; Hamilton, S. K.; Dekker, T.; Garcia, M. H.; Hassan, J. S. *Oil-Particle Interactions and Submergence from Crude Oil Spills in Marine and Freshwater Environments - Review of the Science and Future Science Needs*; Open-File Report 2015-1076; U.S. Geological Survey: Reston, VA, 2015; (b) King, T. L.; Robinson, B.; Boufadel, M.; Lee, K., Flume Tank Studies to Elucidate the Fate and Behavior of Diluted Bitumen Spilled at Sea. *Mar. Pollut. Bull.* **2014,** *83* (1), 32-37; (c) Witt O'Brien's; Polaris Applied Sciences; Western Canada Marine Response Corporation *A Study of Fate and Behavior of Diluted Bitumen Oils on Marine Waters: Dilbit Experiments - Gainford, Alberta*; Trans Mountain Pipeline ULC: 2013; p 163.

10. (a) King, T. L.; Robinson, B.; McIntyre, C.; Toole, P.; Ryan, S.; Saleh, F.; Boufadel, M.; Lee, K., Fate of Surface Spills of Cold Lake Blend Diluted Bitumen Treated with Dispersant and Mineral Fines in a Wave Tank. *Environ. Eng. Sci.* **2015,** *32* (3), 250-261; (b) SL Ross Environmental Research Limited *Meso-scale Weathering of Cold Lake Bitumen/Condensate Blend*; Ottawa, Canada, 2012.

11. Enbridge Energy Partners LP Form 10-Q for the Period Ending June 30, 2014. http://www.sec.gov/Archives/edgar/data/880285/000119312514290178/d765165d10q1.pdf (accessed 10/7/2015).

12. Dollhopf, R. J.; Fitzpatrick, F. A.; Kimble, J. W.; Capone, D. M.; Graan, T. P.; Zelt, R. B.; Johnson, R., Response to Heavy, Non-Floating Oil Spilled in a Great Lakes River Environment: A Multiple-Lines-Of-Evidence Approach for Submerged Oil Assessment and Recovery. In *Proceedings of the International Oil Spill Conference*, Savannah, GA, 2014; pp 434-448.

13. U.S. Energy Information Administration U.S. Imports by Country of Origin - All Countries. http://www.eia.gov/dnav/pet/pet_move_impcus_d_nus_Z00_mbbl_a.htm (accessed 06/11/2015).

14. U.S. Energy Information Administration Refinery Receipts of Crude Oil by Method of Transportation. http://www.eia.gov/dnav/pet/pet_pnp_caprec_dcu_nus_a.htm (accessed 06/11/2015).

15. U.S. Energy Information Administration U.S. Crude Oil Production to 2025: Updated Projection of Crude Types. http://www.eia.gov/analysis/petroleum/crudetypes/pdf/crudetypes.pdf (accessed 10/7/2015).

16. U.S. Energy Information Administration California Field Production of Crude Oil. http://www.eia.gov/dnav/pet/hist/LeafHandler.ashx?n=PET&s=MCRFPCA1&f=M (accessed 06/11/2015).

17. Sheridan, M. *California Crude Oil Production and Imports*; CEC-600-2006-006; California Energy Commission: Sacramento, CA, 2006.

18. Seelke, C. R.; Villarreal, M. A.; Ratner, M.; Brown, P. *Mexico's Oil and Gas Sector: Background, Reform Efforts, and Implications for the United States*; R43313; Congressional Research Service: 2015.

19. National Energy Board Estimated Canadian Crude Oil Exports by Type and Destination. http://www.neb-one.gc.ca/nrg/sttstc/crdlndptrlmprdct/stt/stmtdcndncrdlx-prttpdstn-eng.html (accessed 06/11/2015).

20. National Energy Board Canadian Crude Oil Exports - By Export Transportation System Summary - 5 year trend. http://www.neb-one.gc.ca/nrg/sttstc/crdlndptrlmprdct/stt/cndncrdlxprttrnsprttnsstm5yr/2013/cndncrdlxprttrnsprttnsstm5yr2013-eng.html (accessed 06/11/2015).

21. Adams, J.; Larter, S.; Bennett, B.; Huang, H.; Westrich, J.; C. van Kruisdijk, The Dynamic Interplay of Oil Mixing, Charge Timing, and Biodegradation in Forming The Alberta Oil Sands: Insights from Geologic Modeling and Biogeochemistry. In *Heavy-Oil and Oil-Sand Petroleum Systems in Alberta and Beyond*, Hein, F. J.; Leckie, D.; Larter, S.; Suter, J. R., Eds. American Association of Petroleum Geologists, Canadian Heavy Oil Association, and American Association of Petroleum Geologists Energy Minerals Division: Tulsa, OK, 2013; pp 23-102.

22. Hollebone, B., The Oil Properties Data Appendix. In *Handbook of Oil Spill Science and Technology*, Fingas, M., Ed. John Wiley and Sons Inc.: NY, 2015; pp 577-681.

23. Swarthout, R. F.; Nelson, R. K.; Reddy, C. M.; Hall, C. G.; Boufadel, M.; Valentine, D.; Arey, J. S.; Gros, J., Physical and Chemical Characterization of Canadian Dilbit and Related Products. 2015.

24. Mullins, O. C., The Asphaltenes. *Annu. Rev. Anal. Chem.* **2011,** *4*, 393-418.

25. McKenna, A. M.; Donald, L. J.; Fitzsimmons, J. E.; Juyal, P.; Spicer, V.; Standing, K. G.; Marshall, A. G.; Rodgers, R. P., Heavy Petroleum Composition. 3. Asphaltene Aggregation. *Energy Fuels* **2013,** *27* (3), 1246-1256.

26. Yang, C.; Wang, Z.; Hollebone, B. P.; Brown, C. E.; Yang, Z.; Landriault, M., Chromatographic Fingerprinting Analysis of Crude Oils and Petroleum Products. In *Handbook of Oil Spill Science and Technology*, John Wiley & Sons, Inc: 2014; pp 93-163.

27. Wang, Z.; Hollebone, B. P.; Fingas, M.; Fieldhouse, B.; Sigouin, L.; Landriault, M.; Smith, P.; Noonan, J.; Thouin, G. *Characteristics of Spilled Oils, Fuels, and Petroleum Products: 1. Composition and Properties of Selected Oils*; Environment Canada: 2003.

28. Hollebone, B; Brown, C., Cold Lake Bitumen PAH Analysis Results. Environment Canada ETC Spills Technology Databases, Oil Properties Database.

29. National Research Council, *Oil in the Sea III: Inputs, Fates, and Effects.* The National Academies Press: Washington, DC, 2003.

30. Jokuty, P.; Whiticar, S.; Fingas, M.; Meyer, E.; Knobel, C., Hydrocarbon Groups and Their Relationship to Oil Properties and Behavior. In *Proceedings of the 18th Arctic and Marine Oilspill Program (AMOP) Technical Seminar*, Environment Canada: Ottawa, Canada, 1995; pp 1-19.

31. Environment Canada ETC Spills Technology Databases, Oil Properties Database. http://www.etc-cte.ec.gc.ca/databases/oilproperties/ (accessed 10/7/2015).

32. Aeppli, C.; Carmichael, C. A.; Nelson, R. K.; Lemkau, K. L.; Graham, W. M.; Redmond, M. C.; Valentine, D. L.; Reddy, C. M., Oil Weathering after the Deepwater Horizon Disaster Led to the Formation of Oxygenated Residues. *Environ. Sci. Technol.* **2012,** *46* (16), 8799-8807.

33. (a) Garrett, R. M.; Pickering, I. J.; Haith, C. E.; Prince, R. C., Photooxidation of Crude Oils. *Environ. Sci. Technol.* **1998,** *32* (23), 3719-3723; (b) Maki, H.; Sasaki, T.; Harayama, S., Photo-oxidation of biodegraded crude oil and toxicity of the photo-oxidized products. *Chemosphere* **2001,** *44* (5), 1145-1151; (c) Prince, R. C.; Garrett, R. M.; Bare, R. E.; Grossman, M. J.; Townsend, T.; Suflita, J. M.; Lee, K.; Owens, E. H.; Sergy, G. A.; Braddock, J. F., The roles of photooxidation and biodegradation in long-term weathering of crude and heavy fuel oils. *Spill Sci. Technol. Bull.* **2003,** *8* (2), 145-156; (d) Radović, J. R.; Aeppli, C.; Nelson, R. K.; Jimenez, N.; Reddy, C. M.; Bayona, J. M.; Albaigés, J., Assessment of Photochemical Processes in Marine Oil Spill Fingerprinting. *Mar. Pollut. Bull.* **2014,** *79* (1–2), 268-277.

34. D'Auria, M.; Emanuele, L.; Racioppi, R.; Velluzzi, V., Photochemical Degradation of Crude Oil: Comparison Between Direct Irradiation, Photocatalysis, and Photocatalysis on Zeolite. *J. Hazard. Mater.* **2009,** *164* (1), 32-38.

35. Chapelle, F., *Ground-water microbiology and geochemistry.* John Wiley & Sons: 2001.

36. Barron, M. G.; Carls, M. G.; Short, J. W.; Rice, S. D., Photoenhanced Toxicity of Aqueous Phase and Chemically Dispersed Weathered Alaska North Slope Crude Oil to Pacific Herring Eggs and Larvae. *Environ. Toxicol. Chem.* **2003**, *22* (3), 650-660.

37. (a) Prince, R. C., Petroleum Spill Bioremediation in Marine Environments. *Crit. Rev. Microbiol.* **1993**, *19* (4), 217-242; (b) Transportation Safety Board of Canada *Crude Oil Pipeline - Third-Party Damage*; Pipeline Investigation Report P07H0040; Trans Mountain Pipeline L.P.: Burnaby, Canada, 2007.

38. (a) Boufadel, M. C.; Sharifi, Y.; Van Aken, B.; Wrenn, B. A.; Lee, K., Nutrient and Oxygen Concentrations within the Sediments of an Alaskan Beach Polluted with the *Exxon Valdez* Oil Spill. *Environ. Sci. Technol.* **2010**, *44* (19), 7418-7424; (b) Li, H.; Boufadel, M. C., Long-Term Persistence of Oil from the Exxon Valdez Spill in Two-Layer Beaches. *Nat. Geosci.* **2010**, *3* (2), 96-99.

39. (a) Geng, X.; Boufadel, M. C.; Personna, Y. R.; Lee, K.; Tsao, D.; Demicco, E. D., BIOB: a mathematical model for the biodegradation of low solubility hydrocarbons. *Mar. Pollut. Bull.* **2014**, *83* (1), 138-147; (b) Torlapati, J.; Boufadel, M. C., Evaluation of the Biodegradation of Alaska North Slope Oil in Microcosms Using the Biodegradation Model BIOB. *Front. Microbiol.* **2014**, *5*, 212.

40. Wang, Z.; Fingas, M., Separation and Characterization of Petroleum Hydrocarbons and Surfactant in Orimulsion Dispersion Samples. *Environ. Sci. Technol.* **1996**, *30* (11), 3351-3361.

41. U.S. Environmental Protection Agency *Environmental Response Team's Final Bench Scale/Screening Level Oil Biodegradation Study*; Report Number 1597; 2013.

42. (a) Hamoda, M. F.; Hamam, S. E. M.; Shaban, H. I., Volatilization of Crude Oil from Saline Water. *Oil Chem. Pollut.* **1989**, *5* (5), 321-331; (b) Stiver, W.; Mackay, D., Evaporation Rate of Spills of Hydrocarbons and Petroleum Mixtures. *Environ. Sci. Technol.* **1984**, *18* (11), 834-840.

43. Fingas, M. F., Modeling Oil and Petroleum Evaporation. *J. Pet. Sci. Res.* **2013**, *2* (3), 104-115.

44. (a) Camilli, R.; Reddy, C. M.; Yoerger, D. R.; Van Mooy, B. A. S.; Jakuba, M. V.; Kinsey, J. C.; McIntyre, C. P.; Sylva, S. P.; Maloney, J. V., Tracking Hydrocarbon Plume Transport and Biodegradation at Deepwater Horizon. *Science* **2010**, *330* (6001), 201-204; (b) Reddy, C. M.; Arey, J. S.; Seewald, J. S.; Sylva, S. P.; Lemkau, K. L.; Nelson, R. K.; Carmichael, C. A.; McIntyre, C. P.; Fenwick, J.; Ventura, G. T.; Van Mooy, B. A. S.; Camilli, R., Composition and Fate of Gas and Oil Released to the Water Column During the Deepwater Horizon Oil Spill. *Proc. Natl. Acad. Sci. U. S. A.* **2012**, *109* (50), 20229-20234; (c) Ryerson, T. B.; Camilli, R.; Kessler, J. D.; Kujawinski, E. B.; Reddy, C. M.; Valentine, D. L.; Atlas, E.; Blake, D. R.; de Gouw, J.; Meinardi, S.; Parrish, D. D.; Peischl, J.; Seewald, J. S.; Warneke, C., Chemical Data Quantify *Deepwater Horizon* Hydrocarbon Flow Rate and Environmental Distribution. *Proc. Natl. Acad. Sci. U. S. A.* **2012**, *109* (50), 20246-20253.

45. (a) Fay, J. A., The Spread of Oil Slicks on a Calm Sea. **1971**, 53-63; (b) Hoult, D. P., Oil Spreading on the Sea. *Annu. Rev. Fluid Mech.* **1972**, *4*, 341-368.

46. (a) National Research Council, *Oil Spill Dispersants: Efficacy and Effects*. The National Academies Press: Washington, DC, 2005; (b) Zhao, L.; Boufadel, M. C.; Adams, E. E.; Socolofsky, S. A.; Lee, K., A Numerical Model for Oil Droplet Evolution Emanating from Blowouts. In *Proceedings of the International Oil Spill Conference*, Savannah, GA, 2014; pp 561-571.

47. Johansen, Ø.; Brandvik, P. J.; Farooq, U., Droplet Breakup in Subsea Oil Releases – Part 2: Predictions of Droplet Size Distributions with and without Injection of Chemical Dispersants. *Mar. Pollut. Bull.* **2013**, *73* (1), 327-335.

48. Boufadel, M. C.; Abdollahi-Nasab, A.; Geng, X.; Galt, J.; Torlapati, J., Simulation of the Landfall of the *Deepwater Horizon* Oil on the Shorelines of the Gulf of Mexico. *Environ. Sci. Technol.* **2014**, *48* (16), 9496-9505.

49. Valentine, D. L.; Fisher, G. B.; Bagby, S. C.; Nelson, R. K.; Reddy, C. M.; Sylva, S. P.; Woo, M. A., Fallout Plume of Submerged Oil from *Deepwater Horizon. Proc. Natl. Acad. Sci. U. S. A.* **2014,** *111* (45), 15906-15911.

50. (a) Boufadel, M. C.; Bechtel, R. D.; Weaver, J., The Movement of Oil Under Non-Breaking Waves. *Mar. Pollut. Bull.* **2006,** *52* (9), 1056-1065; (b) Boufadel, M. C.; Du, K.; Kaku, V.; Weaver, J., Lagrangian Simulation of Oil Droplets Transport Due to Regular Waves. *Environ. Modell. Softw.* **2007,** *22* (7), 978-986.

51. Fingas, M.; Fieldhouse, B., Studies on Crude Oil and Petroleum Product Emulsions: Water Resolution and Rheology. *Colloids Surf., A* **2009,** *333* (1-3), 67-81.

52. National Transportation Safety Board *Enbridge Incorporated Hazardous Liquid Pipeline Rupture and Release, Marshall, Michigan, July 25, 2010*; NTSB/PAR-12/01; Washington, DC, 2012.

53. Zhao, L.; Torlapati, J.; Boufadel, M. C.; King, T.; Robinson, B.; Lee, K., VDROP: A Comprehensive Model for Droplet Formation of Oils and Gases in Liquids - Incorporation of the Interfacial Tension and Droplet Viscosity. *Chem. Eng. J.* **2014,** *253* (1), 93-106.

54. (a) Frelichowska, J.; Bolzinger, M. A.; Chevalier, Y., Effects of solid particle content on properties of o/w Pickering emulsions. *J. Colloid Interface Sci.* **2010,** *351* (2), 348-356; (b) Le Floch, S.; Guyomarch, J.; Merlin, F. X.; Stoffyn-Egli, P.; Dixon, J.; Lee, K., The Influence of Salinity on Oil-Mineral Aggregate Formation. *Spill Sci. Technol. Bull.* **2002,** *8* (1), 65-71.

55. Stoffyn-Egli, P.; Lee, K., Formation and Characterization of Oil–Mineral Aggregates. *Spill Sci. Technol. Bull.* **2002,** *8* (1), 31-44.

56. Fitzpatrick, F. A.; Boufadel, M. C.; Johnson, R.; Lee, K.; Graan, T. P.; Bejarano, A. C.; Zhu, Z.; Waterman, D.; Capone, D. M.; Hayter, E.; Hamilton, S. K.; Dekker, T.; Garcia, M. H.; Hassan, J. S. *Oil-Particle Interactions and Submergence from Crude Oil Spills in Marine and Freshwater Environments—Review of the Science and Future Science Needs*; Open-File Report 2015–1076; U.S. Geological Survey: 2015.

57. Lee, K.; Bugden, J.; Cobanli, S.; King, T.; McIntyre, C.; Robinson, B.; Ryan, S.; Wohlgeschaffen, G. *UV-Epifluorescence Microscopy Analysis of Sediments Recovered from the Kalamazoo River*; Centre for Offshore Oil, Gas and Energy Research (COOGER): Dartmouth, Nova Scotia, 2012.

58. (a) Khelifa, A.; Hill, P. S.; Lee, K., The Role of Oil-Sediment Aggregation in Dispersion and Biodegradation of Spilled Oil. In *Oil Pollution and its Environmental Impact in the Arabian Gulf Region*, Al-Azab, M.; El-Shorbagy, W.; Al-Ghais, S., Eds. Elsevier: Amsterdam, Netherlands, 2005; pp 131-145; (b) Lee, K., Oil-Particle Interactions in Aquatic Environments: Influence on The Transport, Fate, Effect and Remediation of Oil Spills. *Spill Sci. Technol. Bull.* **2002,** *8* (1), 3-8; (c) Sun, J.; Khelifa, A.; Zheng, X.; Wang, Z.; So, L. L.; Wong, S.; Yang, C.; Fieldhouse, B., A Laboratory Study on the Kinetics of the Formation of Oil-Suspended Particulate Matter Aggregates Using the Nist-1941b Sediment. *Mar. Pollut. Bull.* **2010,** *60* (10), 1701-1707.

59. Wang, C. Y.; Calabrese, R. V., Drop Breakup in Turbulent Stirred-Tank Contactors. Part II: Relative Influence of Viscosity and Interfacial Tension. *AIChE J.* **1986,** *32* (4), 667-676.

60. Perez, S.; Furlan, P.; Hussein, N.; Shinn, D.; Crook, R., Interaction Between Oil and Suspended Sediments in Class 1-2 Rivers (Poster). In *Proceedings of the International Oil Spill Conference*, Savannah, GA, 2014; p 299120.

61. Waterman, D. M.; Garcia, M. H. *Laboratory Tests of Oil-Particle Interactions in a Freshwater Riverine Environment with Cold Lake Blend Weathered Bitumen*; No. 106; University of Illinois: Urbana, Illinois, 2015.

62. Short, J. W. *Susceptibility of Diluted Bitumen Products from the Alberta Tar Sands to Sinking in Water*; A51148; JWS Consulting LLC: 2013.

63. Fetter, C. W., *Contaminant Hydrogeology*. 2nd ed.; Prentice Hall: Upper Saddle River, NJ, 1999.

64. Ng, G. H. C.; Bekins, B. A.; Cozzarelli, I. M.; Baedecker, M. J.; Bennett, P. C.; Amos, R. T., A Mass Balance Approach to Investigating Geochemical Controls on Secondary Water Quality Impacts at a Crude Oil Spill Site Near Bemidji, MN. *J. Contam. Hydrol.* **2014,** *164,* 1-15.

65. Wesley, J. K. *Kalamazoo River Assessment*; Special Report 35; Michigan Department of Natural Resources, Fisheries Division: Ann Arbor, MI, 2005.

66. (a) Fitzpatrick, F. A.; Johnson, R.; Zhu, Z.; Waterman, D.; McCulloch, R. D.; Hayer, E. J.; Garcia, M. H.; Boufadel, M.; Dekker, T.; Hassan, J. S.; Soong, D. T.; Hoard, C. J.; Lee, K., Integrated Modeling Approach for Fate and Transport of Submerged Oil and Oil-Partricle Aggregates in a Freshwater Riverine Environment. In *Proceedings of the Joint Federal Interagency Conference on Sedimentation and Hydrologic Modeling*, Reno, NV, 2015; (b) Soong, D. T.; Hoard, C. J.; Fitzpatrick, F. A.; Zelt, R. B., Preliminary Analysis of Suspended Sediment Rating Curves for the Kalamazoo River and its Tributaries from Marshall to Kalamazoo, Michigan. In *Proceedings of the Joint Federal Interagency Conference on Sedimentation and Hydrologic Modeling*, Reno, NV, 2015.

67. Hult, M. F. *Ground-Water Contamination by Crude Oil at the Bemidji, Minnesota, Research Site: U.S. Geological Survey Toxic Waste--Ground-Water Contamination Study*; Report 84-4188; 1984.

68. Essaid, H. I.; Bekins, B. A.; Herkelrath, W. N.; Delin, G., N., Crude Oil at the Bemidji Site - 25 Years of Monitoring, Modeling, and Understanding. *Groundwater* **2011,** *49* (5), 706-726.

69. Delin, G. N.; Essaid, H. I.; Cozzarelli, I. M.; Lahvis, M. H.; Bekins, B. A. *Ground Water Contamination by Crude Oil near Bemidji, Minnesota*; USGS Fact Sheet 084-98; U.S. Geological Survey: Mounds View, MN, 1998.

70. Michigan Petroleum Pipeline Task Force *Michigan Petroleum Pipeline Task Force Report*; Michigan Department of Attorney General Lansing, MI, 2015.

71. Michel, J.; Rutherford, N. *Oil Spills in Marshes: Planning & Response Considerations*; U.S. Department of Commerce, National Oceanic and Atmospheric Administration, and the American Petroleum Institute: Washington, DC, 2013.

72. Shen, H. W.; Julien, P. Y., Erosion and Sediment Transport. In *The Handbook of Hydrology*, Maidment, D. R., Ed. McGraw-Hill: New York, 1993.

73. Geng, X.; Boufadel, M. C., Impacts of evaporation on subsurface flow and salt accumulation in a tidally influenced beach. *Water Resour. Res.* **2015,** *51* (7), 5547-5565.

74. Crude Quality Inc. CrudeMonitor.ca. http://www.crudemonitor.ca/ (accessed 10/6/2015).

75. 3 Companies Plead Guilty to Burnaby Oil Spill. *CBC News* October 3, 2011.

76. Shang, D.; Buday, C.; van Aggelen, G.; Colodey, A., Toxicity Evaluation of the Oil Surface Washing Agent Corexit® 9580 and its Shoreline Application in Burrard Inlet, British Columbia. In *Proceedings of the 35th Arctic and Marine Oilspill Program (AMOP) Technical Seminar*, Environment Canada: Ottawa, Canada, 2012.

77. Stantec *Summary of Clean up and Effects of the 2007 Spill of Oil from Trans Mountain Pipeline to Burrard Inlet*; Project No. 1231-10505; 2012.

78. Madison, B. N.; Hodson, P. V.; Langlois, V. S., Diluted Bitumen Causes Deformities and Molecular Responses Indicative of Oxidative Stress in Japanese Medaka Embryos. *Aquat. Toxicol.* **2015,** *165,* 222-230.

79. (a) Colavecchia, M. V.; Backus, S. M.; Hodson, P. V.; Parrott, J. L., Toxicity of Oil Sands to Early Life Stages of Fathead Minnows (Pimephales Promelas). *Environ. Toxicol. Chem.* **2004**, *23* (7), 1709-1718; (b) Colavecchia, M. V.; Hodson, P. V.; Parrott, J. L., CYP1A Induction and Blue Sac Disease in Early Life Stages of White Suckers (Catostomus commersoni) Exposed to Oil Sands. *J. Toxicol. Environ. Health, Part A* **2006**, *69* (10), 967-994; (c) Colavecchia, M. V.; Hodson, P. V.; Parrott, J. L., The Relationships among CYP1A Induction, Toxicity, and Eye Pathology in Early Life Stages of Fish Exposed to Oil Sands. *J. Toxicol. Environ. Health, Part A* **2007**, *70* (18), 1542-1555; (d) Tetreault, G. R.; McMaster, M. E.; Dixon, D. G.; Parrott, J. L., Using Reproductive Endpoints in Small Forage Fish Species to Evaluate the Effects of Athabasca Oil Sands Activities. *Environ. Toxicol. Chem.* **2003**, *22* (11), 2775-2782.

80. Papoulias, D. M.; Velez, V.; Nicks, D. K.; Tillitt, D. E. *Health Assessment and Histopathologic Analyses of Fish Collected from the Kalamazoo River, Michigan, Following Discharges of Diluted Bitumen Crude Oil from the Enbridge Line 6B*; Administrative Report; U.S. Geological Survey: Reston, VA, 2014.

81. Agency for Toxic Substances and Disease Registry *Draft Toxicological Profile for Hydrogen Sulfide and Carbonyl Sulfide*. U.S. Department of Health and Human Services: Atlanta, GA, 2014.

82. Committee on Environment and Natural Resources *Interagency Assessment of Oxygenated Fuels*; National Science and Technology Council: Washington, DC, 1997.

83. U.S. Centers for Disease Control and Prevention Drinking Water, Water Sources. http://www.cdc.gov/healthywater/drinking/public/water_sources.html (accessed 11/13/2015).

84. Railroad Commission of Texas. Field Guide for the Assessment and Cleanup of Soil and Groundwater Contaminated with Condensate From a Spill Incident. http://www.rrc.state.tx.us/oil-gas/environmental-cleanup-programs/guidance-documents-and-helpful-links/condensate-spill-guidance/ (accessed 11/13/2015).

85. Michigan Department of Community Health *Public Health Assessment: Kalamazoo River/ Enbridge Spill: Evaluation of crude oil release to Talmadge Creek and Kalamazoo River on residential drinking water wells in nearby communities (Calhoun and Kalamazoo Counties, Michigan)*; U.S. Department of Health and Human Services: Atlanta, GA, 2013.

86. Baker, M. E.; Wiley, M. J.; Seelbach, P. W., GIS-Based Hyirologic Modeling of Riparian Areas: Implications for Stream Water Quality. *J. Am. Water Resour. Assoc.* **2001**, *37* (6), 1615-1628.

87. American Society of Civil Engineers Task Committee on Modeling of Oil Spills, State-of-the-Art Review of Modeling Transport and Fate of Oil Spills. *J. Hydraul. Eng.* **1996**, *122* (11), 594-609.

88. Lambert, P.; Goldthorp, M.; Fieldhouse, B.; Jones, N.; Laforest, S.; Brown, C. E., Health and safety concerns at dilbit crude oil spills for Environment Canada's responders. In *Proceedings of the 38th Arctic and Marine Oilspill Program (AMOP) Technical Seminar*, Environment Canada: Vancouver, Canada, 2015; pp 664-678.

89. Harrill, J. A.; Wnek, S. M.; Pandey, R. B.; Cawthon, D.; Nony, P.; Goad, P. T., Strategies for Assessing Human Health Impacts of Crude Oil Releases. In *Proceedings of the International Oil Spill Conference*, Savannah, GA, 2014; pp 1668-1685.

90. U.S. Environmental Protection Agency Bridger Pipeline Release. http://www2.epa.gov/region8/bridger-pipeline-release (accessed 11/13/2015).

91. National Oceanic and Atmospheric Administration *Shoreline Assessment Manual*; HAZMAT Report No. 2000-1; Seattle, WA, 2013.

92. Whelan, A.; Clark, J.; Andrew, G.; Michel, J.; Benggio, B., Developing Cleanup Endpoints for Inland Oil Spills. In *Proceedings of the International Oil Spill Conference*, Savannah, GA, 2014; pp 1267-1280.

93. Tsaprailis, H. *Properties of Dilbit and Conventional Crude Oils*; 2480002; Alberta Innovates: 2013.

94. Fingas, M., Diluted Bitumen (Dilbit): A Future High Risk Spilled Material. In *Proceedings of Interspill*, Amsterdam, Netherlands, 2015; p 24.

95. Fieldhouse, B.; Mihailov, A.; Moruz, V., Weathering of Diluted Bitumen and Implications to the Effectiveness of Dispersants. In *Proceedings of the 37th Arctic and Marine Oilspill Program (AMOP) Technical Seminar*, Environment Canada: Ottawa, Canada, 2014; pp 338-352.

96. Guenette, C. C.; Sveum, P.; Buist, I.; Aunaas, T.; Godal, L. *In Situ Burning of Water-in-Oil Emulsions*; SINTEF Report STF21 A94053; 1994.

97. Michel, J.; Benggio, B.; Keane, P., Pre-Authorization for The Use of Solidifiers: Results and Lessons Learned. In *Proceedings of the International Oil Spill Conference*, Savannah, GA, 2008; pp 345-348.

98. Brown, H. M.; Goodman, R. H., The Recovery of Spilled Heavy Oil with Fish Netting. In *Proceedings of the International Oil Spill Conference*, Washington, DC, 1989; pp 123-126.

99. Brown, H. M.; Nicholson, P., The Containment of Heavy Oil in Flowing Water. In *Proceedings of the 15th Arctic and Marine Oil Spill Program (AMOP) Technical Seminar*, Edmonton, Canada, 1992; pp 457-465.

100. Michel, J.; Galt, J. A., Conditions under which floating slicks can sink in marine settings. In *Proceedings of the International Oil Spill Conference*, Long Beach, CA, 1995; pp 573-576.

101. McLinn, E. L.; Stolzenberg, T. R., Ebullition-Facilitated Transport of Manufactured Gas Plant Tar from Contaminated Sediment. *Environ. Toxicol. Chem.* **2009**, *28* (11), 2298-2306.

102. Dollhopf, R., Michel, J., Ed. U.S. Environmental Protection Agency: 2015.

103. National Oil and Hazardous Substances Pollution Contingency Plan. Code of Federal Regulations, Title 40, Part 300, 1994.

104. Comprehensive Environmental Response,Compensation, and Liability Act (CERCLA). 42 U.S. Code §9610.

105. Oil Pollution Act of 1990 (OPA). 33 U.S. Code §2701-2761.

106. U.S. Environmental Protection Agency. *Facility Response Planning Compliance Assistance Guide*; Oil Program Center: Washington, DC, 2002.

107. U.S. Environmental Protection Agency. National Response Team. http://www2.epa.gov/emergency-response/national-response-team (accessed 11/13/2015).

108. U.S. Environmental Protection Agency. Regional Response Teams. http://www2.epa.gov/emergency-response/regional-response-teams (accessed 11/13/2015).

109. U.S. Environmental Protection Agency. Area Contingency Planning. http://www2.epa.gov/oil-spills-prevention-and-preparedness-regulations/area-contingency-planning (accessed 11/13/2015).

110. Clean Water Act. 33 U.S. Code §1251 et seq.

111. Facility Response Plans. *Code of Federal Regulations*, Section 112.20, Title 40, 2012.

112. Executive Order 12777. Implementation of Section 311 of the Federal Water Pollution Control Act of October 18, 1972, as Amended, and the Oil Pollution Act of 1990. 1991.

113. Response Plans for Onshore Oil Pipelines. *Code of Federal Regulations*, Section 194, Title 49.

114. Transportation of Hazardous Liquids by Pipeline. *Code of Federal Regulations*, Section 195, Title 49.

115. Gilbride, P.; Barnes-Weaver, E.; Strasser, M. A.; Wake, S. *EPA Could Improve Contingency Planning for Oil and Hazardous Substance Response*; Report Number 13-P-0152; Office of Inspector General: Washington, DC, 2013.

116. (a) Facilities Transferring Oil or Hazardous Material in Bulk. *Code of Federal Regulations*, Section 154, Title 33; (b) Oil or Hazardous Material Pollution Prevention Regulations for Vessels. *Code of Federal Regulations*, Section 155, Title 33; (c) Caplis, J. R. *MER Policy Letter 03-13; Oil Spill Removal Organization (OSRO) Classification Program*; U.S Coast Guard: Washington, DC, 2013.
117. McNutt, M., A Community for Disaster Science. *Science* **2015,** *348* (6230), 11.

Appendix A

Glossary

Acute—An event occurring over a short time, usually a few minutes or hours. An acute effect happens within a short time after exposure. An acute exposure can result in short-term or long-term health effects. Acute toxicity to aquatic organisms is estimated from short exposures, usually 24, 48, or 96 hours and lethality (death) is the typical endpoint. Results from acute toxicity tests usually report the lethal concentration of the toxicant that causes death to 50% of the test organisms (LC50). The lower the LC50 value the greater the toxicity of the toxicant.

Bitumen—A mixture of hydrocarbons that is too viscous to flow under ambient conditions. Commercial quantities are recovered by thermal processes.

Chronic—An event occurring over a long period of time, generally weeks, months, or years. Chronic exposures occur over an extended period of time or over a significant fraction of an organism's lifetime. Chronic toxicity to aquatic organisms can be estimated from partial life-cycle tests of relatively short duration depending on the organism (i.e., 7 – 21 days) and growth and reproduction are the typical endpoints. Results from chronic toxicity tests are reported as the toxicant concentration that causes a given effect.

Commonly transported crude oils—The oils carried by most transmission pipelines in the United States. Available data show that currently >70% of these are light and medium crude oils.

Conventional oil—Oil that is produced by drilling and pumping from a naturally permeable, subsurface reservoir.

Crude oil—Naturally occurring, unrefined petroleum hydrocarbons extracted from the earth to serve as feedstock for the petroleum industry.

Diluted bitumen—Bitumen diluted with lighter hydrocarbons or a combination of light hydrocarbons such as natural-gas condensate, lighter crude oil, or synthetic oil, such that its viscosity is reduced.

Diluted heavy oil—Heavy oil that is diluted with lighter hydrocarbons such as natural-gas condensate, lighter crude oil, or synthetic oil, such that its viscosity is reduced.

Gas condensate—A low-density mixture of hydrocarbon liquids obtained by condensing the less-volatile components of raw natural gas.

Heavy crude oil—A naturally occurring, unrefined petroleum product with a density greater than 0.93 g/cm^3 or an API gravity less than 20.

High molecular weight petroleum compounds—Compounds with a molecular weight greater than about 250 Daltons; typically these compounds are viscous liquids or solids at ambient temperatures.

Light crude oil—A naturally occurring, unrefined petroleum product with a density ranging from 0.80 to 0.85 g/cm^3 or an API gravity ranging from 35 to 45.

Low molecular weight petroleum compounds—Compounds with a molecular weight lower than about 250 Daltons; typically these compounds are liquids at ambient temperatures.

Medium crude oil—A naturally occurring, unrefined petroleum product with a density ranging from 0.85 to 0.92 g/cm^3 or an API gravity ranging from 36 to 21.

Sublethal—Toxicant is below the concentration that directly causes death. Exposures to sublethal concentrations of a toxicant may produce less obvious and measureable effects on behavior, molecular, biochemical, cellular and/or physiological function (e.g., growth and reproduction) and histology of organisms.

Synthetic crude—Oil produced from bitumen by physical and chemical processes less elaborate than those used in full-scale refineries and implemented near the site of production.

Transmission pipeline—A continuous pipe used to transport oil and petroleum products from gathering points to storage or distribution points. It is distinct from smaller, shorter pipelines used to collect oil from individual wells or to distribute to points of consumption.

Unconventional oil—Oil that is produced by unconventional means, including thermal separation of non-liquid bitumen from a host rock and hydraulic fracturing of impermeable reservoirs or source rocks.

Undiluted heavy oil—Heavy oil that is transported without dilution. It may be heated to facilitate transport.

Upgraded bitumen—Bitumen that has been subject to some refinement to remove or convert some of the more recalcitrant components. It is also an intermediate in the production of synthetic crude oil.

Appendix B

Committee Member and Staff Biographies

Committee

Diane McKnight (*chair*) is a professor of civil, environmental and architectural engineering and a fellow of the Institute of Arctic and Alpine Research at the University of Colorado. Her research focuses on interactions between hydrologic, chemical, and biological processes in controlling the dynamics in aquatic ecosystems. This research is carried out through field-scale experiments, modeling, and laboratory characterization of natural substrates. In addition, Dr. McKnight conducts research focusing on interactions between freshwater biota, trace metals, and natural organic material in diverse freshwater environments, including lakes and streams in the Colorado Rocky Mountains and in the McMurdo Dry Valleys in Antarctica. She interacts with state and local groups involved in mine drainage and watershed issues in the Rocky Mountains. Dr. McKnight is a member of the National Research Council's (NRC's) Polar Research Board and is a former member of the Water Science and Technology Board. She is a past president of the American Society of Limnology and Oceanography and the Biogeosciences section of the American Geophysical Union. She received her Ph.D. in environmental engineering from the Massachusetts Institute of Technology and is a member of the National Academy of Engineering.

Michel Boufadel is professor of environmental engineering and director of the Center for Natural Resources Development and Protection at the New Jersey Institute of Technology. He is a professional engineer in

Pennsylvania and New Jersey. Dr. Boufadel has conducted, since 2001, research projects funded by the U.S. Environmental Protection Agency and the National Oceanic and Atmospheric Administration (NOAA) on oil dispersion and transport offshore. He has adopted a multiscale approach where he conducts experiments in flasks and wave tanks of various sizes and models processes from the microscopic scale to the sea scale. Dr. Boufadel was involved in the response to the Deepwater Horizon blowout and assisted NOAA personnel conducting various tasks within the response. Dr. Boufadel has more than 80 refereed articles in publications such as *Marine Pollution Bulletin, Environmental Science and Technology,* and the *Journal of Geophysical Research.* He also has more than 30 publications in oil spill conference proceedings, such as those of the International Oil Spill Conference and Arctic and Marine Oil Spill. He is an associate editor of the *Journal of Environmental Engineering,* American Society of Civil Engineers.

Merv Fingas is a scientist working on oil and chemical spills. He was Chief of the Emergencies Science Division of Environment Canada for over 30 years and is currently working on research in Western Canada. Dr. Fingas has a Ph.D. in environmental physics from McGill University, and three master's degrees—chemistry, business, and mathematics—all from University of Ottawa. He also has a bachelor of science in chemistry from Alberta and a bachelor of arts from Indiana. He has more than 860 papers and publications in the field. Dr. Fingas has prepared seven books on spill topics and is working on two others. He has served on two committees on the National Academy of Sciences of the U.S. on oil spills including the recent "Oil in the Sea." He is chairman of several ASTM and intergovernmental committees on spill matters. Importantly, he was the founding chairman of the ASTM subcommittee on in situ burning and chairman of oil spill treating agents and another on oil spill detection and remote sensing, positions he holds today.

Dr. Fingas began his career in 1974 working for Environment Canada as a scientist working on oil and chemical spills. His first tasks were largely to work on the Beaufort Sea Studies, a multi-million-dollar joint industry-government program to develop oil spill readiness for the Canadian Beaufort Sea. His role in these studies was to coordinate chemical and physical studies and to prepare overview documents. With the completion of these studies a new study of large magnitude, the Arctic and Marine Oil Spill Program, was founded in 1977. Dr. Fingas, one of the founders of this program, worked in general coordination on the program and specifically managed a number of subprojects, including those on chemistry, oil behavior, remote sensing, spill tracking, and spill treating agents. Dr. Fingas continued in many of these research fields until today. His specialties include Arctic oil spills, oil chemistry, spill dynamics and behavior, spill treating agents, remote sensing and detection, spill tracking, and in situ burning. He continues research and writing on these topics to this day.

Stephen Hamilton is currently a professor of ecosystem ecology and biogeochemistry at Michigan State University. He received his Ph.D. at the University of California, Santa Barbara, in 1994. His principal research interests involve ecosystem ecology and biogeochemistry, with particular attention to nutrients and biogeochemical processes in aquatic environments as well as agricultural ecosystems. His research integrates approaches from varied disciplines such as geology, chemistry, remote sensing, and hydrology as well as ecology. He has conducted research on various aspects of aquatic ecosystems in southern Michigan, including wetlands, streams, lakes, and watersheds. He also works on tropical ecosystems in South America and dryland river ecosystems in Australia. Since 2006 he has been President of the Kalamazoo River Watershed Council. He served as an independent, volunteer advisor to the U.S. Environmental Protection Agency for the 2010 Marshall, MI pipeline release of oil sands crude into the Kalamazoo River, and was a member of its Scientific Support Coordination Group.

Orville "O.B." Harris is President of O.B. Harris, LLC, which is an independent consultancy specializing in the regulation, engineering, and planning of petroleum liquids pipelines. Currently, he is the Independent Monitoring Contractor for the Consent Decree between the U.S. and BP Alaska, Inc. From 1995 to 2009, he was Vice President of Longhorn Partners Pipeline, L.P., which operated a 700-mile pipeline that carried gasoline and diesel fuel from Gulf Coast refineries to El Paso, Texas. In this position, he was responsible for engineering, design, construction, and operation of the system. From 1991 to 1995, he was President of ARCO Transportation Alaska Inc., which owned four pipeline systems, including a portion of the Trans Alaska Pipeline System (TAPS) which transports crude oil from the North Slope of Alaska to the Port of Valdez. From 1977 to 1990, he held several supervisory and managerial positions at the ARCO Pipeline Company, including District Manager for Houston and Midland, Texas, Manager of the Northern Area, and Manager of Products Business. While at ARCO Transportation, he directed the efforts of a team of corrosion experts guiding $400 million of repairs to the TAPS system. He is a past member of the Board of Directors of the Association of Oil Pipelines and the Pipeline and Hazardous Materials Safety Administration's Technical Hazardous Liquids Pipeline Safety Standards Committee. Mr. Harris joins this committee having previously served on the committee for the preceding NRC study, *Effects of Diluted Bitumen on Crude Oil Transmission Pipelines.* He holds a bachelor's degree in civil engineering from the University of Texas and an M.B.A. from Texas Southern University.

John Hayes is Scientist Emeritus and retired Director of the National Ocean Sciences Accelerator Mass Spectrometry Facility (NOSAMS) at the Woods Hole Oceanographic Institution. His work has dealt with isotope effects in biochemical reactions and their significance and utility in studies of geochemical processes. Recent topics have included studies of carbon- and hydrogen-isotopic fractionations imposed by phytoplankton and other microorganisms, paleoenvironmental studies based on sedimentary isotopic and organic-geochemical records, studies of the anaerobic oxidation of methane in marine sediments, the long-term record of ^{13}C in sedimentary organic carbon, and developments in stable-isotopic analytical techniques. He has served as chair of the Organic Geochemical Division of the Geochemical Society and on the Executive Committee of the Integrated Ocean Drilling Program. He received his Ph.D. in chemistry from the Massachusetts Institute of Technology and is a member of the National Academy of Sciences and of the American Academy of Arts and Sciences.

Jacqueline Michel is a geochemist specializing in terrestrial and marine pollution studies, coastal geomorphology, and environmental risk assessments. She has specialized expertise in the behavior, tracking, recovery, and effects of submerged oil. Having worked in 32 countries, she has extensive international experience and has worked in many different coastal and marine environments.

Dr. Michel is one of the founders of Research Planning, Inc. and has been President since 2000. She often leads multidisciplinary teams on projects where her problem-solving skills are essential to bringing solutions to complex issues. For example, her work during spill emergencies requires her to rapidly develop consensus and provide decision makers with needed information. Because of her routine scientific support for spills, she has extensive knowledge of and practical experience in pollutant fate, transport, and effect issues. She has been a leader in the development of methods and the conduct of Natural Resource Damage Assessments following spills and groundings. She has taken a lead role in 29 damage assessments for federal and state trustees.

Dr. Michel has been recognized for her achievements through appointments to many respected committees and panels, including four National Academies committees: Spills of Nonfloating Oil (1999); Oil in the Sea (2002); Chair of Spills and Emulsified Fuels: Risk and Response (2001);

and Chair of the Committee on Understanding Oil Spill Dispersants: Efficacy and Effects (2005). She was on the Oceans Board for 2001-2005 and is a Lifetime Associate of the National Academies. She was on the Science Advisory Panel to the U.S. Commission on Ocean Policy. She is an adjunct professor in the School of the Environment, University of South Carolina. She has written over 225 technical publications.

Carys Mitchelmore earned her Ph.D. from the University of Birmingham (UK) in 1997 investigating toxicity processes and effects in aquatic organisms exposed to organic pollutants, including crude oil and its constituent polycyclic aromatic hydrocarbons (PAHs). Dr. Mitchelmore is an associate professor at the University of Maryland Center for Environmental Science, Chesapeake Biological Laboratory, in Solomons, MD. Her expertise lies in aquatic toxicology and her research experience includes understanding routes of exposure, bioaccumulation, metabolism, depuration, trophic transfer and the target sites of pollutants, including PAHs and emerging contaminants of concern. Investigations have used an array of organisms, from bacteria, algae, and invertebrate and vertebrate species, such as oysters, blue crabs, anemones, corals, fish, and reptiles. Current research projects are directed at understanding the uptake, routes of exposure (including chemical partitioning of dissolved and particulate fractions), fate and effects of oil, chemical dispersants (e.g., Corexit and alternatives) and dispersed oil. Focused areas of impact include DNA damage, oxidative stress and antioxidant responses, endocrine disruption, and immune function. Recent studies have investigated the use of oil rig-fouling organisms as biomonitoring tools to provide baseline datasets that can provide essential information regarding the recovery of organisms following a pollution event. Dr. Mitchelmore is also co-author of the 2005 NRC report *Oil Spill Dispersants: Efficacy and Effects* and also provided testimonies to various Senate and House committees following the Deepwater Horizon incident regarding dispersant use. Dr. Mitchelmore is also actively involved in determining the efficacy of various shipboard scale ballast water treatments and in investigating the occurrence and toxicity of chlorinated and brominated organic compounds.

Denise Reed, Ph.D., is a nationally and internationally recognized expert in coastal marsh sustainability and the role of human activities in modifying coastal systems. She has studied coastal issues in the United States and around the world for over 30 years.

Dr. Reed has worked closely with Louisiana's state government in developing coastal restoration plans. Her experience includes helping monitor natural resources in the Pontchartrain Basin following the Deepwater Horizon oil spill in 2010 and researching ecosystem restoration and planning in the California Bay-Delta. She has served on numerous boards and panels addressing the effects of human alterations on coastal environments and the role of science in guiding restoration, including a number of National Research Council committees. Prior to joining The Water Institute of the Gulf, Dr. Reed served as Director of the Pontchartrain Institute for Environmental Sciences and as a professor in the University of New Orleans' Department of Earth and Environmental Sciences. She is a member of the Chief of Engineers Environmental Advisory Board and the Ecosystems Sciences and Management Working Group of the NOAA Science Advisory Board. She earned a bachelor's and doctoral degree in geography from the University of Cambridge.

Robert (Bob) Sussman is the principal in Sussman and Associates, a consulting firm that offers advice and support on energy and environmental policy issues to clients in the nonprofit and private sectors. He is on the adjunct faculty at Georgetown Law Center and Yale Law School. Sussman served for four and a half years in the Obama Administration, first as co-head of the transition team for the Environmental Protection Agency (USEPA) and then as Senior Policy Counsel to the USEPA Administrator. Mr. Sussman previously served in the Clinton Administration as the USEPA Deputy Administrator during 1993-1994. In this position, he was the Agency's Chief Operating Officer and Regulatory Policy Officer.

At the end of 2007, he retired as a partner at the law firm of Latham & Watkins, where he headed the firm's environmental practice in Washington, DC, for 10 years. Previously, he was a partner at Covington & Burling. For several years, Sussman was named one of the leading environmental lawyers in Washington, DC, by Chambers USA: America's Leading Business Lawyers and The International Who's Who of Environmental Lawyers. He was a Senior Fellow at the Center for American Progress in 2008, writing and speaking about climate change and energy. Sussman is a magna cum laude 1969 graduate of Yale College and a 1973 graduate of Yale Law School, where he was an editor of the Yale Law Journal. He clerked for Judge Walter K. Stapleton of the Third Circuit Court of Appeals.

David Valentine currently serves as a professor in the Department of Earth Science at the University of California, Santa Barbara. His research interests focus on the interface of geochemistry and microbiology. Valentine is an Aldo Leopold Leadership Fellow and the recipient of a CAREER award in chemical oceanography from the National Science Foundation. He is best known for his research on archaeal ecology, methane biogeochemistry, oil seeps, and the aftermath of the Deepwater Horizon event, as well as for engagement with popular media."

Staff

Douglas Friedman is a Senior Program Officer with the Board on Chemical Sciences and Technology at the National Academy of Sciences, Engineering, and Medicine in Washington, DC. His primary scientific interests lie in the fields of organic chemistry, organic & bio-organic materials, chemical & biological sensing, and nanotechnology, particularly as they apply to national and homeland security. Dr. Friedman has supported a diverse array of activities since joining the Academies. He served as study director on *Industrialization of Biology: A Roadmap to Accelerate the Advanced Manufacturing of Chemicals; Safe Science: Promoting a Culture of Safety in Academic Chemical Research; Transforming Glycoscience: A Roadmap for the Future; Determining Core Capabilities in Chemical and Biological Defense Science and Technology; Effects of Diluted Bitumen on Crude Oil Transmission Pipelines*; and *Responding to Capability Surprise: A Strategy for U.S. Naval Forces*. Additionally, he has supported activities on *Convergence: Safeguarding Technology in the Bioeconomy, The Role of the Chemical Sciences in Finding Alternatives to Critical Resources; Opportunities and Obstacles in Large-Scale Biomass Utilization*; and *Technological Challenges in Antibiotics Discovery and Development*. Dr. Friedman is currently directing studies on the environmental effects of diluted bitumen oil spills, safeguarding technology in the bioeconomy, and the regulation of biotechnology. Prior to joining the NRC he performed research in physical organic chemistry and chemical biology at Northwestern University, the University of California, Los Angeles, the University of California, Berkeley, and Solulink Biosciences. He received a Ph.D. in Chemistry from Northwestern University and a B.S. in Chemical Biology from the University of California, Berkeley.

Camly Tran joined the Board on Chemical Sciences and Technology at the National Academy of Sciences, Engineering, and Medicine in 2014 as a postdoctoral fellow after receiving her Ph.D. in chemistry from the Department of Chemistry at Brown University and is currently an Associate Program Officer. During her time at Brown, she received various honors including the Elaine Chase Award for Leadership and Service, American Chemical Society Global Research Exchanges Education Training Program, and the Rhode Island NASA grant. Dr. Tran completed the workshop summary *Mesoscale Chemistry* and is currently supporting activities on the environmental effects of diluted bitumen oil spills, the changing landscape of hydrocarbon feedstocks for chemical production, and the standard operating procedures for safe and secure handling, management, and storage of chemicals in chemical laboratories.